श्री:

Vedic Mathematics

(30 Formulae elucidated with simple examples)

*

Dr. Ramamurthy N

M.Sc., B.G.L., C.A.I.I.B., C.C.P., D.S.A.D.P., C.I.S.A., P.M.P., CGBL, Ph.D.

*

Name of the Book:	**Vedic Mathematics** (30 formulae elucidated with simple examples)
First Edition:	2013
Second Edition:	2016
Author:	Dr. Ramamurthy N. http://ramamurthy.jaagruti.co.in

Number of Pages:	147

*****

TABLE OF CONTENTS

*****

Dedication

Mātru Devo Bhava

I dedicate this book with *pranāms* to my mother *Śreemati* **Alankāravalli Ammal**, who brought me to this level right from scratch. She is the Divine Mother *Śree Devee* in the human form in this world. Though she is not physically available to see me at this glassy, I am sure all these are possible only with her blessings and she continues to live with me spiritually.

Dr. Ramamurthy N

*****

Swami Paramarthananda
Disciple of Swami Dayananda Saraswathi
Sriram Apartments,
80, St. Mary's Road
Chennai 600 018.

Foreword

When we study the ancient history of India we discover that there are many things that we can be proud of. While ancient India had a well-developed civilisation, its contribution to the development of human thought was stupendous. These contributions are not only in the field of spiritual sciences, but also in the field of secular sciences like medicine, mathematics, astronomy, politics, economics, linguistics, dramatics, dance music, fine arts, chemistry, architecture, etc. Without modern facilities, instruments, technology, etc., what they have achieved is indeed astounding. Unfortunately, most of these are recorded in Samskrit language and, therefore, inaccessible to the layman. However, a lot of research is being undertaken now-a-days and they are brought out for the appreciation of general public.

In the field of Mathematics the works of Aryabhatta I (5th Century C.E.), Brahmagupta (7th Century C.E.), Aryabhatta II (10th Century C.E.) and Bhaskaracharya (12th Century C.E.) are acknowledged by many scholars of East and West.

In this series of contributions by Indians to the world of Mathematics "Vedic Mathematics" is the latest addition. It is a book written by HH Swami Bharati Krishna Teerthaji and wasfirst published in 1965. It contains a list mental calculation techniques claimed to be based on the ancient Vedas. The mental calculation system mentioned in the book is popularly known by the name "Vedic Mathematics". Although the book was first published in 1965, Teerthaji had been propagating the technqiues since much earlier, through lectures and classes.

Vedic Mathematics is just around 50 years old and has lot of potential and wealth in it. With more and more research in this subject, quiet a lot of complicated calculations/ applications can be made simple and easy.

This book of Sri **Ramamurthy** can be treated as a commentary for the Sutras of Vedic Mathematics written by Teerthaji. In other books on *Vedic* Mathematics, different types of problems are considered and solutions provided using one or more of the formulae or corollaries. However, in this book each of the formulae has been individually discussed and explained with example(s) in simple language so that it could be easily understood by any Mathematics student. The criticisms and replies to the the criticisms are also dealt with alongwith some Management Perspective of Vedic Mathematics. Thus Sri **Ramamurthy** has done a commendable work through this book.

It is also understood that Sri Ramamurthy has written other books relating to Religion, Information Technology, Banking, etc. May he succeed in all his endeavours. I pray God for his long life and continuous service in spreading the knowledge.

With blessings to all the readers.

Chennai
June 2013 *Swami Paramarthananda*

*****

श्रीः

Introduction

Mathematics is the basis of all Sciences. Indian Mathematics is a universe in itself. Once stepped in, anyone will be steeped (submerged) into it. It is very difficult to decide which direction to pursue, which concept to pick up, etc. We have heard of the glorious achievements of traditional Indian thinkers, Astronomers and Philosophers. We also know about the work of such intellectuals in the field of Mathematics. But most of us were not much aware that Mathematics was so common that it was used in the day–to–day activities of the common people. It is this simplicity of Indian Mathematics that fascinated everyone. Some of the basic techniques of Indian and *Vedic* Mathematics in various fields including the fairly recent field of Computer Science are really astonishing.

Vedic Mathematics is relatively new, around 50 years old and hence has lot of potential for study and elaboration. Having delved into Saṁskṛit, anyone will be wonder struck with the huge treasure of knowledge hidden in our tradition.

Vedic Mathematics compiled by late HH *Śaṇkarācārya* (Bhārati Krṣṇa Tīrtha Swāmi) of Govardhan *Pīṭha* is a monumental work. In his deep–layered explorations of cryptic *Vedic* mysteries relating especially to the calculations of laconic formulae and their neat and ready application to practical problems, the late *Śaṇkarācārya* shows the rare combination of probing insight of revealing intuition of a *Yogi* with the analytic acumen and synthetic talent of a Mathematician.

Śrī Bhārati Kṛṣṇa Tīrtha Swāmiji did not write *Vedic* Mathematics on his own. He got it in his mind through penance during later 1910s and gave us the treasure. This was known to us only in early 60's of 20th century C.E., i.e. nearly 50 years after his compiling the same.

The sixteen formulae of *Vedic* Mathematics correspond to sixteen vowels of Saṁskṛit language. With vowels, all the words are formed and without which no word can be found. Similarly any problem of Mathematics can be solved using one or more of the *Vedic* Mathematics formulae. *Vedic* Mathematics is itself called as "mental Mathematics". This gives lot of work to the brain and the mind – leading to various thinking processes. *Vedic* formulae not only tell us how to do the Mathematical calculations by easy one line method and through rapid processes, but they also tabulate the results in the shape of special corollaries containing merely illustrative specimens with a master–key for "unlocking other portals" too.

Vedic Mathematics is not only a sophisticated pedagogic and research tool but also an introduction to an ancient civilisation. It takes us back to many millennia of India's Mathematical heritage. Rooted in the ancient *Vedic* sources which heralded the dawn of human history and illumined by their erudite exegesis, India's intellectual, scientific and aesthetic vitality blossomed and triumphed not in only in philosophy, physics, Astronomy, ecology and performing arts but also in Geometry, Algebra and arithmetic.

It is understandable that because of some of the undernoted reasons students are scary and keep a distance from *Vedic* Mathematics:

- One does not have equally good knowledge both in Mathematics and Saṁskṛit as well.
- It is so cryptic that decoding needs lot of thought and application of mind.
- It does not have a uniform method for the same type of calculations – for instance, in the case of normal multiplication we have tables to use and for all the multiplication problems we use the same tables. However, in the case of *Vedic* Mathematics different formulae are used for the numbers ending with '9' or '5', etc.

Vedic Mathematics also triggers the Management Science angle that stimulate the thought of "differently thinking", "diagonally thinking", "vertically thinking", etc. All these thought processes are needed for effective Management. All these are used widely and with a purpose in computer Project/ Program Management also.
Normally it is seen, in any standard text book on *Vedic* Mathematics, different types of problems are considered and solutions provided using one or more of the formulae or corollaries. However, each of the formulae/ corollary has been individually explained with example(s). This makes this book a unique of its kind.

Vedic Mathematics is just around 50 years old and has lot of potential and wealth in it. With more and more excavation in this subject, quiet a lot of complicated calculations/ applications can be made simple and easy. We can with confidence believe that if we look at *Vedic* Mathematics more closely enough, there is still many a thing left to learn and explore.

Conventions used in this book: The transliterated Samskrit words are written in italics – for instance *Vedic*. When Samskrit words are transliterated in English diacritical marks are used to correctly pronounce the words. However the same has **not** been used in its entirety in this book, since it makes the reading a little more difficult and since this book is intended for common audience.

My humble pranams to HH Swami Paramarthananda, who has written the foreword to this edition.

Sincere thanks are due to all those who supported in this noble cause.

The readers are requested to give all comments and feedback to the author.

Om Tat Sat

Chennai
November 2013

Dr. *Ramamurthy N*

*****

1. Vedic Mathematics

The *Vedas* are the most ancient record of human experience and knowledge, passed down orally for generations. It is believed that this should have been written thousands of years ago. Medicine, Architecture, Astronomy and many other branches of knowledge, including Mathematics, are dealt with in these texts. It is not surprising that the country credited with introducing the current number system and the invention of perhaps the most important Mathematical symbol '0', may have more to offer in the field of Mathematics.

Vedic Mathematics, compiled by, **HH Jagadguru Swāmi Śrī Bhārati Kṛṣṇa Tīrthāji Mahāraj (जगद्गुरु स्वामि श्री भारती कृष्ण तीर्थाजी महराज**: March, 1884 – February 2, 1960) of *Govardhan Pīṭha* is a monumental work. The phrase "**compiled by**" and not "written by" is consciously used, for the obvious reason that this was available in *Vedas* and *Śaṅkarācārya* discovered the formulae and did not invent them. In his deep–layer explorations of cryptic *Vedic* mysteries relating especially to the calculus of shorthand formulae and their neat and ready application to practical problems, the *Śaṅkarācārya* shows the rare combination of the probing insight of revealing intuition of a *Yogi* with the analytic acumen and synthetic talent of a Mathematician. A brief biography of the *Śaṅkarācārya* is given at the end of this book.

With the *Śaṅkarācārya*, we belong to a race, now fast becoming extinct, of diehard believers who think that the *Vedas* represent an inexhaustible mine of profoundest wisdom in matters, which are both spiritual and temporal; and that this store of wisdom was not, as regards its assets of fundamental validity and value at least, gathered by the laborious inductive and deductive methods of ordinary systemic enquiry, but was direct gift of revelation to seers and sages who in their higher reaches of *Yogic* realisation were competent to receive it from a source, perfect and immaculate.

Vedic Mathematics offers a fresh and highly efficient approach to Mathematics covering a wide range – right from elementary multiplication till relatively advanced topics like the solution of non–linear partial differential equations. But the *Vedic* scheme is not simply a collection of rapid methods; it is a system, a unified approach. The essence of the system is that it is based on sixteen *sūtra*–s – a *sūtra* is a terse statement of an important point or principle.

Whether or not the *Vedas* are believed as repositories of perfect wisdom, it is unquestionable that the *Vedic* race lived not merely as a pastoral folk possessing a partially developed culture and civilisation. The *Vedic* seers proved themselves adepts in all levels and branches of knowledge, both theoretical and practical. For example, they had their varied objective as science both pure and applied.

Let us consider a concrete illustration: Suppose in a time of drought we require rains by artificial means. The modern scientists have their own theory and art (technique) for producing the result. The old seer scientists also had these both, but different from the form now available. They had science and technique, called *Yajna*, in which *Mantra, Yantra* and other factors must co–operate with Mathematical determinates and precision. For this purpose, they had developed the six auxiliaries of the *Vedas* in each of which the Mathematical skill and adroitness, occult or otherwise, play a decisive role. The *sūtra*–s lay down the shortest and surest lines. The correct intonation of the *Mantra*, the correct configuration of the *Yantra* (in the making of the *Veda*, etc., e.g. the quadrate of a circle), the correct time or astral conjunction factor, the correct rhythms, etc., are all important aspects to be noted. All had to be perfected so as to produce the desired results effectively and adequately. Each of these required precise Mathematical calculations. The modern technician has his logarithmic tables and mechanic's manuals, the old *Yajnik* had their *Sūtra*–s.

The revered *Śaṅkarācārya* used to say that he had reconstructed the sixteen Mathematical formulae from the *Atharva Veda* after assiduous research and *Tapas* (austerity/ penance) for about eight years in the forests surrounding Śringeri. Obviously these formulae are not to be found in the present editions of *Atharva Veda*. They were actually reconstructed, on the basis of intuitive revelation, from materials scattered here and there in the *Atharva Veda*. According to him these aphorisms are contained in the *Parisiṣṭa* (the appendix portions) of the *Atharva Veda*.

1.1. Salient Features of *Vedic* Mathematics

Vedic Mathematics is becoming increasingly popular as more and more people are introduced to the beautifully unified and easy calculation methods. Some interesting salient features in the very words of HH Jagadguru Śri Bhārati Kṛṣṇa Tīrthaji Mahāraj from the preface of his book on "*Vedic* Mathematics" are worth mentioning:

"*Vedas*, the oldest 'Religious' scriptures of the whole world, claim to deal with all branches of learning whether spritual or temporal and to give the earnest seeker after knowledge all the requisite instructions and guidance in full detail and on scientifically – nay – Mathematically – accurate lines in them all and so on".

The very word *Veda* has the derivational meaning, i.e. the fountain–head and illimitable store–house of all knowledge. This derivation, in effect, means, connotes and implies that the *Vedas* should contain within themselves all knowledge needed by mankind. That is, it connotes and implies that our ancient Indian *Vedic* lore should be all round complete, perfect and be able to throw light on all matters which any aspiring seeker after knowledge can possibly seek to be enlightened on.

1. The *Sūtra*–s (aphorisms) apply to almost every branch of Mathematics.
2. The *Sūtra*–s are easy to understand, easy to apply and easy to remember and the whole work can be truthfully summarised in one word as 'Mental'.
3. Even as regards complex problems involving a good number of Mathematical operations (consecutively or even simultaneously to be performed), the time taken by the *Vedic* method will be a third, a fourth, a tenth, or even a much smaller fraction of the time required according to modern (i.e. current) methods.
4. And in some very important and striking cases, sums requiring 30, 50, 100 or even more numerous and cumbrous 'steps' of working (according to the current methods) can be answered in a single and simple step of work by the *Vedic* method. Children (of only 10 or 12 years of age) merely look at the sums written on the blackboard and immediately shout out and convey the answers. This is because, as a matter of fact, each digit automatically yields its predecessor and its successor. And the children have merely to go on tossing off (or reeling off) the digits one

after another (forwards or backwards) by mere mental arithmetic (without needing pen or pencil, paper, slate, etc.).

5. On seeing this kind of work actually being performed by children, the doctors, professors and other 'big–guns' of Mathematics are wonder–struck and exclaim: Is this Mathematics or magic? And invariably the answer is: it is both. It is magic until we understand and it is Mathematics thereafter.
6. As regards the time required by the students for mastering the whole course of *Vedic* Mathematics as applied to all its branches, what is needed merely is, to state from the actual experience, 8 months (or 12 months) at an average rate of 2 or 3 hours per day should suffice for completing the Mathematical studies on these *Vedic* lines.
7. Tough Mathematical problems (which the most advanced present day Mathematics need to spent huge amount of time, energy and money involving large numbers of difficult, tedious and cumbersome 'steps' of working) can easily and readily be solved with the help of these ultra–easy *Vedic Sūtra*–s in a few simple steps and methods. Hence this can conscientiously be described as "mental Mathematics".
8. It is thus in the fitness of things that the four *Vedas* – *Ṛg*, *Sāma*, *Yajus* and *Atharva* respectively include *upa–Vedas* as below, respectively:

I. *Āyur Veda* (medical science).
II. *Dhanur Veda* (archery and other military sciences).
III. *Gāndharva Veda* (the science of art and music) and
IV. *Sthapatya Veda* (engineering, architecture etc. and all branches of Mathematics in general).

All these subjects, it may be noted, are inherent parts of the *Vedas* i.e., are reckoned as 'spiritual' studies.

9. Similar is the case with *Vedāngas* (i.e., grammar, prosody, Astronomy, lexicography etc.) which according to the Indian cultural conceptions are also inherent parts and subjects of *Vedic* (i.e. religious) study.

1.2. Formulae and Corollaries of *Vedic* Mathematics

Vedic Mathematics has sixteen main formulae (*Sūtra*–s) and fourteen corollaries (*Upa–sūtra*–s or *Upa–Sūtra*–s). While there is no second opinion as to the number of formulae, the number and name of the corollaries differ in different schools. However, fourteen corollaries as detailed below have been accepted by majority of the schools. They are listed in the table below:

Vedic Mathematics Formulae

#	*Sūtra*	Formula in English	Translation
The main *Sūtra*–s – main formulae			
1.	एकाधिकेन पूर्वेण	*Ekādhikena Pūrveṇa*	By one more than the previous one
2.	निखिलं नवतश्चरमं दशतः	*Nikhilam Navathaścaramam Dhaśataḥ*	All from 9 and the last from 10.
3.	ऊर्ध्वतिर्यग्भ्याम्	*Ūrdhva Tiryagbhyām*	Vertically and Cross–wise
4.	परावर्त्य योजयेत्	*Parāvartya Yojayet*	Transpose and Apply
5.	शून्यं साम्यसमुच्चये	*Śūnyam Sāmyasamuccaye*	If the *Samuccaye* is the same it is Zero
6.	(आनुरूप्ये) शून्यमन्यत्	*(Ānurūpye) Śūnyamanyat*	If One is in Ratio the Other is Zero
7.	संकलनव्यवकलनाभ्याम्	*Sankalana–vyavakalanābhyām*	By Addition and by Subtraction
8.	पूरणापूरणाभ्याम्	*Pūraṇāpūraṇābhyām*	By the Completion or Non–Completion
9.	चलनकलनाभ्याम्	*Calana–kalanābhyām*	Differential Calculus
10.	यावदूनम्	*Yāvadūnam*	By the Deficiency
11.	व्यष्टिसमष्टिः	*Vyaṣṭisamaṣṭiḥ*	Specific and General
12.	शेषाण्यङ्केन चरमेण	*Śeṣāṇyaṅkena Carameṇa*	The Remainders by the Last digit
13.	सोपान्त्यद्वयमन्त्यम्	*Sopāntyadvayamantyam*	The Ultimate and Twice the Penultimate
14.	एकन्यूनेन पूर्वेण	*Ekanyūnena Pūrveṇa*	By One Less than the One Before

#	*Sūtra*	Formula in English	Translation
15.	गुणितसमुच्चय:	*Guṇitasamuccayaḥ*	The Product of the Sums
16.	गुणकसमुच्चय:	*Guṇakasamuccayaḥ*	All the Multipliers
		The *Sub* or *Upa–Sūtra*–s (corollaries)	
1.	आनुरूप्येण	*Ānurūpyeṇa*	Proportionately
2.	शिष्यते शेषसंज्ञ:	*Śiṣyate Śeṣasamjñaḥ*	The Remainder Remains Constant
3.	आद्यमाद्येन अन्त्यमन्त्येन	*Ādyamādyena Antyamantyena*	The First by the First and the Last by the Last
4.	केवलै: सप्तकं गुण्यात्	*Kevalaiḥ Saptakam Guṇyāt*	For 7 the Multiplicand is 143
5.	वेष्टनम्	*Veṣṭanam*	By Osculation
6.	यावदूनं तावदूनम्	*Yāvadūnam Tāvadūnam*	Lessen by the Deficiency
7.	यावदूनं तावदूनीकृत्य वर्गंच योजयेत्	*Yāvadūnam Tāvadūníkṛtya Vargañca Yojayet*	Whatever the Deficiency lessen by that amount and set up the Square of the Deficiency
8.	अन्त्ययोर्दशकेऽपि	*Antyayordaśake'pi*	Last Totaling 10
9.	अन्त्ययोरेव	*Antyayoreva*	Only the Last Terms
10.	समुच्चयगुणित:	*Samuccayaguṇitaḥ*	The Sum of the Products
11.	लोपनस्थापनाभ्याम्	*Lopanasthāpanābhyām*	By Alternative Elimination and Retention
12.	विलोकनम्	*Vilokanam*	By mere observation
13.	गुणितसमुच्चय: समुच्चयगुणित:	*Guṇitasamuccayaḥ Samuccayaguṇitaḥ*	The Product of the Sum is the Sum of the Products
14.	ध्वजाङ	*Dvajāḍa*	On the flag

The *sūtra*–s embody laws, principles or methods of working and do not always easily succumb to rigid classification. Some of them have many applications. **Transpose and apply** is one such formula. It applies to solving equations, division in fractions and dividing numbers close to a base. It has many other uses at higher stages in Mathematics and indicates, not a single or particular algorithm, but a general mental procedure. There are other formulae, such as, **when the final digits add up to ten**, for which the uses appear to be very limited. It is because of the very few formulae which have many faceted qualities, that the subject becomes greatly unified and simplified.

More than one application may be used to solve a single Mathematical problem. Hence it is generally not possible to clearly list down which formula is to be used for a specific problem. However, a generic list of the applications of the above *sūtra*–s is given at the end of this book.

Normally corollaries are derived from the main theorems. But here, it can be observed that the *Upa–Sūtra*–s are not used under the main *sūtra*–s. They at times themselves act as individual *sūtra*–s. This is the main deviation from the usual Mathematical principles of theorem (*sūtra*) and corollaries (*Upa–Sūtra*).

The main reasons for this deviation could be:

- In general Mathematics – corollaries are for theorems and not for formulae.
- For want of better word we have translated the *upa–Sūtra*–s as corollary but in actual sense it is not a true meaning of that word and English is found wanting in this case for a better word to convey the exact meaning.

Actually a theorem or formula is not the exact English equivalent of the word '*sūtra*', for the word *sūtra* conveys more meanings like a laconic expression, when expanded yields many a meaning. But however for convenience *sūtra* continues to be called as formula.

Applications:

In many a practical area *Vedic* Mathematics formulae are being applied and exploited to the utmost benefit. Some of them are indicated here. This is only a small sample and much more to list.

The computer processors using the algorithms based on Indian/ *Vedic* Mathematics have been proved to be much faster than those using the normal algorithms. They are not only faster, but use very less resources like memory, sharing of time to other terminals, etc. It is a double benefit that time is saved and resultantly the cost is also reduced. It has been evidenced that Indian/ *Vedic* Mathematics can also be used in modern day electronic systems including Computers, which use algorithms in processors, micro–processors, mini–processors, co–processors, etc.

Digital signal processing (DSP) is the technology that is omnipresent in almost every engineering discipline. Faster additions and multiplications are of extreme importance in DSP for convolution, discrete Fourier transform, digital filters, etc. The core computing process is always a multiplication routine; therefore, DSP engineers are constantly looking for new algorithms and hardware to implement them. The *Vedic* mathematics formulae, manifesting a unified structure of Mathematics, are highly useful in this regard.

Vedic Mathematics formulae are being applied in furtherance of Pythagoras theorem, Pythagorean Triplets, etc.

*****

2. Sūtra 1: एकाधिकेन पूर्वेण – Ekādhikena Pūrveṇa

– By one more than the previous one

Eka – one; *Adhika* – more; and *Pūrva* – before

The *Sūtra* reads: एकाधिकेन पूर्वेण – *Ekādhikena Pūrveṇa*, which translated into English, simply means "By one more than the previous one". Its application and modus operandi are:

The preposition 'by' (in the *Sūtra*) indicates that the arithmetical operation prescribed is either multiplication or division. For, in the case of addition or subtraction, to or from respectively would have been the appropriate proposition to use. The inference is therefore obvious that either multiplication or division must be enjoined. Again both the meanings are perfectly correct and equally tenable, according to grammar and literary usage. Hence, since there is no reason – in or from the text – for only one of the meanings being accepted and the other one rejected, it follows that both the processes are actually meant.

Squares of numbers ending in 5:

Let us consider an example 65^2

For the number 65, the last digit is 5 and the 'previous' digit is 6.

Hence, 'one more than the previous one', that is, 6 + 1 gives us 7.

The first part is the first number, 6, multiplied by the number "one more", .i.e. 7.

This becomes the left part of the result, that is, 6 x 7= 42.

Since the number is ending with 5, the right part is always 25.

Thus 65^2 = 6 x 7 / 25 = 4225.

Vulgar fractions:

One is absolute and this absolute is the innermost Self of us all. Everything comes from 'One' and without it nothing could be made. Since One is absolute, it is unchanging – forever the same. As such it cannot be divided. The number One is indivisible.

In the creation we pretend that One can be divided. We pretend that it can be divided into two, three, four and so on.

If two numbers are written as a fraction one below the other with a line in between then it is called vulgar fraction. It can also be expressed as a decimal fraction with a decimal point in between.

Vulgar fractions whose denominators are numbers ending with NINE:

Ekādhikena process can be effectively used both in division and multiplication.

a) Division Method: To find the value of $1 \div 19$.

The number of recurring decimal places is the difference of numerator and denominator,

i.e., $19 - 1 = 18$ places.

For the denominator 19, the *Pūrva* (previous) is 1.

Hence *Ekādhikena Pūrva* is $1 + 1 = 2$.

Now the division method follows:

Step 1: Divide the numerator 1 by 20.
$1 \div 20 = 0.1 \div 2 = .{}_{1}0$ (0 times, 1 remainder)

Step 2: Divide 10 by 2 – $0.0{}_{0}5$ (5 times, 0 remainder)
Step 3: Divide 5 by 2 – $0.05{}_{1}2$ (2 times, 1 remainder)

Step 4: Divide ${}_{1}2$ i.e., 12 by 2 – 0.0526 (6 times, No remainder)

Step 5: Divide 6 by 2 – 0.05263 (3 times, No remainder)

Step 6: Divide 3 by 2 – $0.05263{}_{1}1$ (1 time, 1 remainder)

Step 7: Divide $_{1}1$ i.e. 11 by 2 – $0.052631_{1}5$ (5 times, 1 remainder)

Step 8: Divide $_{1}5$ i.e. 15 by 2 – $0.0526315_{1}7$ (7 times, 1 remainder)

Step 9: Divide $_{1}7$ i.e., 17 by 2 – $0.05263157\ _{1}8$ (8 times, 1 remainder)

Step 10: Divide $_{1}8$ i.e., 18 by 2 – 0.0526315789 (9 times, No remainder)

Step 11: Divide 9 by 2 – $0.0526315789\ _{1}4$ (4 times, 1 remainder)

Step 12: Divide $_{1}4$ i.e., 14 by 2 – 0.052631578947 (7 times, No remainder)

Step 13: Divide 7 by 2 – $0.052631578947_{1}3$ (3 times, 1 remainder)

Step 14: Divide $_{1}3$ i.e. 13 by 2 – $0.0526315789473_{1}6$ (6 times, 1 remainder)

Step 15: Divide $_{1}6$ i.e. 16 by 2 – 0.052631578947368 (8 times, No remainder)

Step 16: Divide 8 by 2 – 0.0526315789473684 (4 times, No remainder)

Step 17: Divide 4 by 2 – 0.05263157894736842 (2 times, No remainder)

Step 18: Divide 2 by 2 – 0.052631578947368421 (1 time, No remainder)

Now from step 19, i.e., dividing 1 by 2, Step 2 to Step 18 repeats thus giving

1 ÷ 19 = 0.052631578947368421 or 0.052631578947368421

It can be noted that we have completed the process of division only by using '2'. Nowhere the division by 19 occurs.

b) **Multiplication Method:** To find the value of 1 ÷ 19

First we recognise that the last digit of the denominator of the type 1 ÷ 19 as 9.

For 1 ÷ 19, 'previous' digit is 1 and one more than it is 1 + 1 = 2.

Therefore 2 is the multiplier for the conversion.

We write the last digit in the numerator as 1 and follow the steps leftwards.

Step 1: 1

Step 2: 21(multiply 1 by 2, put to left)

Step 3: 421(multiply 2 by 2, put to left)

Step 4: 8421(multiply 4 by 2, put to left)

Step 5: $_{1}68421$ (multiply 8 by 2 =16, 1 carried over, 6 put to left)

Step 6: $_{1}368421$ (6 x 2 = 12 + 1 [carry over] = 13, 1 carried over, 3 placed to left)

Step 7: 7368421 (3 x 2, = 6 +1 [carryover] = 7, put to left)

Step 8: $_{1}47368421$ (as in the same process)

Step 9: 947368421 (continue to step 18)

Step 10: $_{1}8947368421$

Step 11: $_{1}78947368421$

Step 12: $_{1}578947368421$

Step 13: 1578947368421

Step 14: 31578947368421

Step 15: 631578947368421

Step 16: $_{1}2631578947368421$

Step 17: 52631578947368421

Step 18: 1052631578947368421

Now from step 18 onwards the same numbers and order towards left continue.

Thus 1 ÷ 19 = 0.052631578947368421

Now, if we multiply the last digit 0 by 2 and add the carry 1, we get the partial sum as 1, which is the digit where we started. Proceeding with the above process will lead only in the repetition of the above sequence. Hence we stop here. The entire calculation involves simple multiplication and addition and can be done mentally.

The generic method for finding the reciprocal of any number n, which is of the form 10x + 9, is as follows:

1. The last digit of the recurring set of numbers of the reciprocal is 1
2. Initially the carry is 0
3. Multiply the latest digit of the result by (x + 1) and add with the previous carry to get a partial sum.
4. Divide the partial sum by 10. Write the remainder as the latest digit of the result and mark the quotient as the carry.

5. Repeat steps 3 and 4 till the partial sum becomes 1.

In the given example, the last digit is 9. The previous digit is 1. One more than the previous digit is 2. In accordance with the *sūtra*, we multiply it by 2. Similarly the multiplier for 49 is 5, for 149 is 15, for 12789 is 1279 and so on.

Another shortcut: If it is keenly observed, in the reciprocal of 19, the recurring part has 18 digits. It can be seen that the 18th digit and the 9th digit added together gives 9. Similarly, the sum of 17th and 8th digits, 16th and 7th digits etc., gives 9. Thus by finding the last 9 digits of the result, we can get the remaining 9 digits by merely subtracting from 9. The workload has become half.

*****

3. Sūtra 2: निखिलं नवतश्चरमं दशतः – *Nikhilam Navataścaramam Daśataḥ*

– All from 9 and the last from 10.

Nikhilam – All, complete, whole, entire, full; *Navataḥ* – Nine; *Carama* – the last, final, end, outermost; and *Daśataḥ* – Ten

The *Sūtra* reads: निखिलं नवतश्चरमं दशतः – *Nikhilam Navataścaramam Daśataḥ*, which translated into English, simply reads "All from 9 and the last from 10". Its application and modus operandi are:

The *Sūtra* basically means; start from the left most digits and begin subtracting '9' from each of the digits; but subtract '10' from the last digit.

The *Sūtra* can be used to perform instant subtractions for numbers consisting of 1 followed by zeroes like 100, 1000, 10,000, etc. Basically, we start using the first zero in the number and then subtract 9. In the case of a number in which we have more 'zeros' than figures, like say 1000 – 76, take it as 076 and apply the same rule.

Example 1: Let us choose the number 6. This has only one digit; hence it is also the last digit. Thus applying the *Nikhilam Sūtra* we have 10 subtracted from 6 to get '–4'.

Example 2: Given the number 87, it is clear that the first digit is 8 and the last digit is 7.

Using the *Sūtra*: Subtract 9 from 8 to get '–1';
Subtract 10 from the last digit 7 to get '–3'.
Hence on the application of the *Nikhilam Sūtra* we get '–13'.

Example 3: Subtract 4679 from 10000

```
1 0 0 0 0
  ↓ ↓ ↓ ↓
  4 6 7 9
  -------
  5 3 2 1
  -------
```

Nikhilam Application: Multiplication

Multiplication: When any number is multiplied by one there is no change. It is only when the multiplier is two or more that there can be any increase. Philosophically this can be indicated through the book of Genesis, where God created male and female, that is two, before there could be any multiplication.

We all know that a complement is that which relates a number to unity. In Mathematics the unity is expressed as 1 or 1 followed by any number of zeroes.

For example the complement of the number 86 is 14
I.e. We take all from nine and the last from ten:

8 from 9 = 1
6 from 10 = 4 and hence 14.

Example 4: To multiply 92 and 89. Apply *Nikhilam Sūtra* – "All from nine and last from ten" on both the Numbers. Write this down side by side.

92 –08
89 –11

Multiply (–08) and (–11) to get '88'

92 –08
89 –11
88

Now we cross–add. This is done by either adding 92 and –11 or 89 and –08 to get '81'.

92 –08

89 –11

It can be noted that in both operations we get the same answer that is '81' which is written as below to get the solution.

92 –08
89 –11

81 88
Hence 92 × 89 = 8188.

Example 5: What if the multiplicands are not near any convenient base?

Let us use an example to explain 42 x 46

For this purpose we use an *Upa–Sūtra*, which simply states *'Ānurūpyeṇa'* – Proportionately. In practical terms it means that we can choose our working base, a multiple or sub–multiple of the power of ten, as long as we multiply or divide the left hand side of the result accordingly.

Let the working base be 50 = (100/ 2)

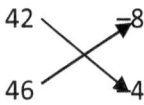

42 –8
46 4

38 (/2)/ 32 = 19 / 32

As we see, we treat the left hand side to the same operation we treated to the base – division by 2.

We could have also taken 40 as our working base and 10 as our base (40) = 10 x 4

42 +2
46 +6
48 (x 4) / (1)2 = 192 / (1)2 = 193 / 2

Example 6: What if one number is above and the other is below the base?

Let us use an example to explain 108 x 92

Let us take base as 100

108 + 8
x 94 − 6

---

102 / (48)

---

101 / 52 (since it is less than the base, the compliment of 48 has to be taken and −1 is carried over)

Hence the result is 10152

Nikhilam division

At '1' there is no division. Division always starts at 2. In the book of Genesis, there is division on the very first day of creation. The division into two at the beginning of creation is also the division into good and evil.

द्वौ भूतसर्गौ लोकेऽस्मिन् दैव आसुर एव च
dvou bhūtasargou loke'smin daiva āsura eva ca (*Bhagavad Gītā* 16-6)

There are two types of created beings in this world, the divine and demoniacal (Good and the Evil).
Some examples of certain applications of *Nikhilam* division:

Example 7: 53 divided by 9.

The first digit 5 is the quotient. Now add the first and second digit: 5 + 3 = 8.

The remainder is 8.

If the first and second digits add up to more than nine, subtract 9 from the remainder and add 1 to the quotient.

Example 8: 69 divided by 9.

6 + 9 = 15.

Subtract 9 from 15 to get the real remainder (6) and add one to the quotient. Thus, the quotient is 7 and the remainder is 6.

Example 9: 1234 divided by 9

- The first digit of the quotient will be the same as the first digit of the dividend: 1
- Add the first two digits (1 + 2 = 3) and place that after the first digit (1). Now we have 13.
- Add the first three digits (1 + 2 + 3 = 6) and place that after 13. Now we have 136. This is the quotient.
- The remainder will be all for digits added together. 1 + 2 + 3 + 4 = 10
- Since the remainder is more than 9, subtract 9 from the remainder and add one to the quotient.
- The quotient is 137 and the remainder is 1.

For divisors other than 9 the *Nikhilam Sūtra* all from 9 and the last from 10 is used. For 9, subtracting 9 from 10, we get 1 as the multiplier.

Consider 23 ÷ 9 – The tens digit 2 is the quotient. Multiply it by 1 (multiplier) to get 2, 2 + 3 = 5, the remainder.

Consider 23 ÷ 8. The tens digit 2 is the quotient.

Multiplier in this case is 10 – 8 = 2.

Hence remainder is 2 (first digit) x 2 (Multiplier) + 3 (second digit) = 7.

*****

4. Sūtra 3: ऊर्ध्वतिर्यग्भ्याम् – *Ūrdhvatiryagbhyām*

– Vertically and Cross–wise

Ūrdhvam – vertically; *Tiryag* – horizontally; and *Bhyām* – use both.

The *Sūtra* reads: ऊर्ध्वतिर्यग्भ्याम् – *Ūrdhvatiryagbhyām* which translated into English, simply says "Vertically and Cross–wise". This is one of the most used formulae. This is the general formula applicable to all cases of multiplication and also in the division of a large number by another large number. Its application and modus operandi are:

(a) Multiplication of two 2 digit numbers.

Example 1: Find the product 14 x 12. Symbolically we can represent the process as follows:

Step i):

1 4 ↓
1 2

: 4 x 2

Step ii):

1 4 (cross-wise)
1 2

2 + 4 : 8

Step iii):

↓ 1 4
1 2

1 x 1 : 6 : 8 = 168

What happens when one of the results i.e., either in the last digit or in the middle digit of the result, contains more than 1 digit?

Answer is simple. The right most digit thereof is to be put down there and the preceding, i.e., left hand side digit or digits should be carried over to the left and placed under the previous digit or digits of the upper row. The digits carried over may be written as in example 2 below:

Example 2: 32 x 24

Step (i):

3 2

2 4

2 x 4 = 8

Step (ii):

3 2

2 4

12 + 4 = 16.

Here 6 is to be retained. 1 is to be carried out to left side.

Step (iii):

3 2

2 4

3 x 2 = 6

Now the carried over digit 1 of 16 is to be added. i.e. 6 + 1 = 7.

Thus 32 x 24 = 768

The Above Steps can be easily remembered by the following line diagrams

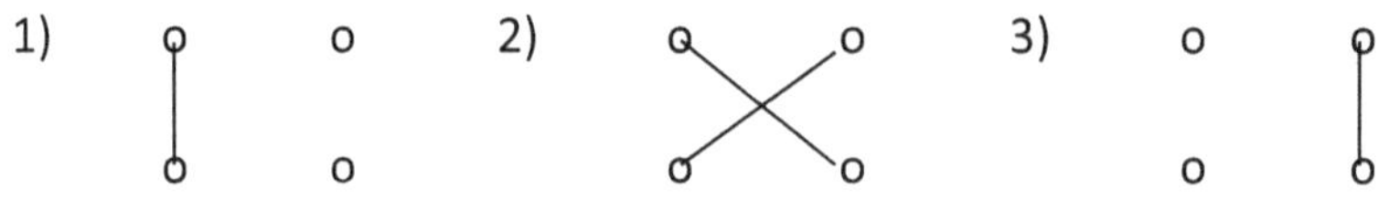

This method can be extended for any number of digits.

For 3 digit numbers, the line diagram can be represented as follows

1) 

2) 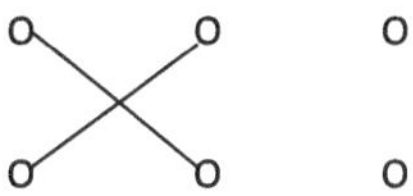

3) 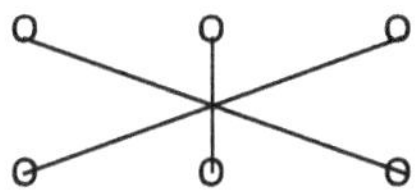

4) 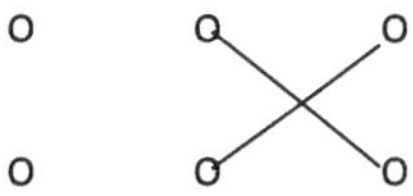

5) 

The Line Diagram for four digits is given below:

1) 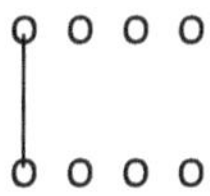

2) 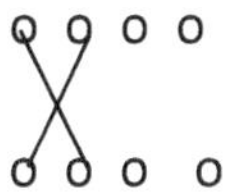

3) 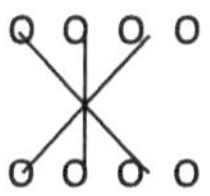

4) 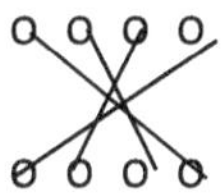

5) 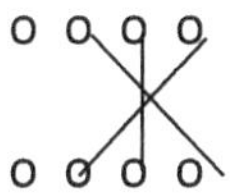

6) 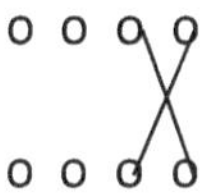

7) 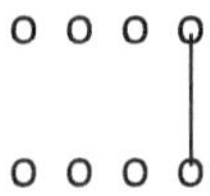

(b) Sums of products –

Example 3: Calculate 213x426 + 354x631 + 652x172

In this case, we have to carry out three "vertically and crosswise" multiplications simultaneously.

	2	1	3
	4	2	6 x
+			
	3	5	4
	6	3	1 x
+			
	6	5	2
	1	7	2 x

$$4\ \ 2_{10}\,6_{12}\,2\ {}_{5}5\ {}_{2}6$$

This formula vertically and crosswise can be used in various Mathematical applications like:

- Evaluation of determinants
- All operations of determinants
- Solving simultaneous linear equations
- Inverse of matrices
- Curve–fitting
- Evaluation of logarithms and exponentials
- Finding Cosine, Sine and Tangents and their inverses.
- Solving of transcendental equations
- Solving Cubic and higher order equations
- Solving of linear and non–linear differential, integral and integro–differential equations.
- Solving of linear and non–linear partial differential equations

Couple of above applications is explained in other chapters/ sub-chapters.

*****

5. Sūtra 4: परावर्त्य योजयेत् – *Parāvartya Yojayet*

– Transpose and Apply

Parāvara – distant and near, earlier and later, highest and lowest, cause and effect, the whole extent of an idea; *Parāvartya* – turned back, exchanged or reversed; and *Yojayet* – connect, join or choose, make use of.

The *Sūtra* reads: परावर्त्य योजयेत् – *Parāvartya Yojayet*, which translated into English, simply says "Transpose and Apply". Its application and modus operandi are:

The well–known rule relating to transposition enjoins invariable change of sign with every change of side. Thus '+' becomes '–' and its converse; and similarly 'x' becomes '÷' and vice versa.

(i) Let us consider the division by divisors of more than one digit and when the divisors are slightly greater than powers of 10.

Example 1: To divide 1225 by 12.

Step 1: From left to right write the divisor. Leaving the first digit, write the other digit or digits using negative (–) sign and place them below the divisor as shown.

```
 12
 -2
----
```

Step 2: Write down the dividend to the right. Set apart the last digit for the remainder.

```
i.e.  12   122   5
      -2
```

Step 3: Write the 1st digit below the horizontal line drawn under the dividend. Multiply the digit by –2, write the product below the 2nd digit and add.

```
i.e.  12   122   5
      –2    –2
    ____________________
             10
```

Since 1 x (–2) = (–2) and 2 + (–2) = 0

Step 4: We get second digits' sum as '0'. Multiply the second digits' sum thus obtained by (–2) and write the product under 3rd digit and add.

```
12      122    5
– 2     –20
----    ----------
        102    5
```

Step 5: Continue the process to the last digit.

```
i.e.  12      122    5
      –2      –20   –4
      -----   ----------
              102    1
```

Step 6: The sum of the last digit is the remainder and the result to its left is Quotient.

Thus Q = 102 and R = 1

Example 2: Divide 1697 by 14.

```
14      1  6  9  7
–4        –4 –8 –4
----    -------
        1  2  1  3
```

Q = 121, R = 3.

Simultaneous simple equations:

By applying *Parāvartya Sūtra* we can derive the values of x and y, which are given by two simultaneous equations. The values of x and y are given by ratio form. The method to find out the numerator and denominator of the ratio is given below;

Example 3: To solve $2x + 3y = 13$
$4x + 5y = 23$

i) To get x, start with coefficients of y and the constant terms and cross–multiply forward, i.e. right-ward. Start from the upper row and multiply across by the lower one and conversely, the connecting link between the two cross–products being a minus. This becomes the numerator.

$$2x + 3y = 13$$
$$4x + 5y = 23$$

Numerator of the x value is $3 \times 23 - 5 \times 13 = 69 - 65 = 4$

ii) Go from the upper row across to the lower one, i.e. the coefficient of x but backward, i.e. left-ward.

Denominator of the value of x value is $3 \times 4 - 2 \times 5 = 12 - 10 = 2$
Hence value of $x = 4 \div 2 = 2$.

iii) To get y, follow the cyclic system, i.e. start with the constant term on the upper row towards the coefficient of x on the lower row.

Hence numerator of the value of y is $13 \times 4 - 23 \times 2 = 52 - 46 = 6$.

iv) The denominator is the same as obtained in Step (ii) i.e. 2.
Hence the value of y is $6 \div 2 = 3$.
Thus the solution to the given equation is $x = 2$ and $y = 3$.

Example 4: To solve $5x - 3y = 11$
$6x - 5y = 09$

Now the Numerator of x is $(-3)(9) - (-5)(11) = -27 + 55 = 28$
Denominator of x is $(-3)(6) - (5)(-5) = -18 + 25 = 07$
$x = 28 \div 7 = 4$
and for y, Numerator is $(11)(6) - (9)(5) = 66 - 45 = 21$
Denominator is 7
Hence $y = 21 \div 7 = 3$

*****

6. *Sūtra* 5: शून्यं साम्यसमुच्चये – *Śūnyam Sāmyasamuccaye*

– If the *Samuccaye* is the same, it is Zero

Samuccayam – aggregation, collection, accumulation; *Sama* – same, similar, equal; and *Śūnyam* – Zero, nothing, naught.

The *Sūtra* reads: शून्यं साम्यसमुच्चये – *Śūnyam Sāmyasamuccaye*, which translated into English, simply says "If the *Samuccaye* is the same, it is Zero". Its application and modus operandi are:

The *Sūtra* – *Śūnyam Sāmyasamuccaye* means – if *Samuccaye* is the same then that *Samuccaye* is Zero. i.e., it should be equated to zero. The term *Samuccaya* has different meanings under different contexts.

i) First we interpret '*sāmyasamuccaya*' as a term which occurs as a common factor in all the terms concerned and proceed as follows.

Example 1: The equation $7x + 3x = 4x + 5x$ has the common factor 'x' in all its terms. Hence by the *Sūtra* it is zero, i.e. $x = 0$

Otherwise we have to work it as:

$$7x + 3x = 4x + 5x$$
$$10x = 9x$$
$$10x - 9x = 0; \quad x = 0$$

This is applicable not only for 'x' but also for any such unknown quantity as follows.

Example 2: $5(x+1) = 3(x+1)$

Now *Samuccaye* is $(x + 1)$

$$x + 1 = 0 \quad \text{gives} \quad x = -1$$

ii) Now we interpret '*Sāmyasamuccaya*' as product of independent terms in expressions like $(x + a)(x + b)$

Example 3: $(x + 3)(x + 4) = (x - 2)(x - 6)$

Here *Samuccaye* is $3 \times 4 = 12 = (-2) \times (-6)$.

Since it is the same, we derive $x = 0$

iii) Here we interpret *'Samuccaya'* as the sum of the denominators of two fractions having the same numerical numerator.

Example 4:

$$\frac{1}{3x-2} + \frac{1}{2x-1} = 0$$

For this we proceed by taking L.C.M.

$$\frac{(2x-1)+(3x-2)}{(3x-2)(2x-1)} = 0$$

$$\frac{(5x-3)}{(3x-2)(2x-1)} = 0$$

$$5x - 3 = 0 \quad \Rightarrow \quad x = 3/5$$

Instead of this, we can directly put the *Sāmuccaye* i.e., sum of the denominators

i.e., $3x - 2 + 2x - 1 = 5x - 3 = 0$ giving $5x = 3 \Rightarrow x = 3/5$

For all problems of the type $(m \neq 0)$

$$\frac{m}{(ax+b)} + \frac{m}{(cx+d)} = 0$$

Samuccaye is $ax + b + cx + d$ and solution is $x = \dfrac{(b+d)}{(a+c)}$

iv) We now interpret *'Samuccaya'* as combination or total.

If the sum of the numerators and the sum of the denominators be the same, then that sum = 0.

Let us consider examples of type

$$\frac{(ax + b)}{(ax + c)} = \frac{(ax + c)}{(ax + b)}$$

As per *Sūnya Samuccaya*, $x = \frac{-(c + b)}{2a}$

Example 5:

$$\frac{(3x + 4)}{(3x + 5)} = \frac{(3x + 5)}{(3x + 4)}$$

Hence from *Sūnya Samuccaye* we get 6x + 9 = 0 or x = –3/2
Consider the examples of the type, where $N_1 + N_2 = K(D_1 + D_2)$, where K is a numerical constant, then also by removing the numerical constant K, we can proceed as above.

Example 6:

$$\frac{(2x + 3)}{(4x + 5)} = \frac{(x + 1)}{(2x + 3)}$$

Here $N_1 + N_2 = 2x + 3 + x + 1 = 3x + 4$
$D_1 + D_2 = 4x + 5 + 2x + 3 = 6x + 8 = 2\,(3x + 4)$

Removing the numerical factor 2, we get 3x + 4 on both sides.

$3x + 4 = 0;\ 3x = -4 \Rightarrow x = -4 / 3.$

v) Solving Quadratic equations.

In this context, we take the problems as follows;

If $N_1 + N_2 = D_1 + D_2$ and $N_1 - D_1 = N_2 - D_2$ then both are equated to zero.

Example 7:

$$\frac{(3x + 2)}{(2x + 5)} = \frac{(2x + 5)}{(3x + 2)}$$

Observe $N_1 + N_2 = 3x + 2 + 2x + 5 = 5x + 7;$

$D_1 + D_2 = 2x + 5 + 3x + 2 = 5x + 7$

Further $N_1 - D_1 = (3x + 2) - (2x + 5) = x - 3;$

$N_2 - D_2 = (2x + 5) - (3x + 2) = - x + 3 = -(x - 3)$

Hence $5x + 7 = 0, x - 3 = 0$

i.e. $x = -7 / 5, x = 3$

Example 8:

$$\frac{1}{x-4} + \frac{1}{x-6} = \frac{1}{x-2} + \frac{1}{x-8}$$

Now *Samuccaya Sūtra*, tells us that, if other elements being equal, the sum–total of the denominators on the L.H.S. and their total on the R.H.S. be the same, that total is zero.

Now $D_1 + D_2 = x - 4 + x - 6 = 2x - 10$ and $D_3 + D_4 = x - 2 + x - 8 = 2x - 10$
By *Samuccaya*, $2x - 10$ gives $2x = 10$; $x=5$

Śūnyam Sāmya Samuccaye **in Certain Cubes:**

If we have to solve the problem $(x - 249)^3 + (x + 247)^3 = 2(x - 1)^3$.

The traditional method will be horrible even to think of.

But *Samuccaya Sūtra* $(x - 249) + (x + 247) = 2x - 2 = 2 (x - 1)$.

And hence $(x - 1)$.

On R.H.S. cube, it is enough to state that $x - 1 = 0$ by the *Sūtra*, $x = 1$ is the solution.

No cubing or any other Mathematical operations need to be performed.

*****

7. *Sūtra* 6: (आनुरूप्ये) शून्यमन्यत् – (*Ānurūpye*) *Śūnyamanyat* – If one is in Ratio the other is Zero

Ānurūpye – conformity, suitableness; *Śūnyam* – Zero, nothing and *Anyat* – Everything else, other things.

The *Sūtra* reads: (आनुरूप्ये) शून्यमन्यत् – (*Ānurūpye*) *Śūnyamanyat*, which translated into English, simply says "if one is in Ratio the other is Zero". Its application and modus operandi are:

We use this *Sūtra* in solving a special type of simultaneous simple equations in which the co–efficients of one variable are in the same ratio to each other as the independent terms are to each other. In such a context the *Sūtra* says the 'other' variable is zero from which we get two simple equations in the first variable and of course give the same value for the variable.

Example 1:

$$3x + 7y = 2$$
$$4x + 21y = 6$$

It may be observed that the coefficients of y are in the ratio 7 : 21 i.e., 1 : 3, which is same as the ratio of constant terms i.e., 2 : 6 i.e., 1 : 3.

Hence the other variable x = 0 and hence 7y = 2 or 21y = 6 gives y = 2 / 7

Example 2:

$$323x + 147y = 1615$$
$$969x + 321y = 4845$$

Using *Ānurūpye Sūtra* we get x = 5, because the ratio of coefficients of x is 323 : 969 = 1 : 3

and the ratio of the constant terms is also 1615 : 4845 = 1 : 3.

Hence y = 0 and hence 323x = 1615 or 969x = 4845 hence x = 5.

In solving simultaneous quadratic equations, we can take the help of the '*Sūtra*' in the following way:

Example 3:

$$x + 4y = 10$$

$$x^2 + 5xy + 4y^2 + 4x - 2y = 20$$

$x^2 + 5xy + 4y^2 + 4x - 2y = 20$ can be written as

$$(x + y)(x + 4y) + 4x - 2y = 20$$

$$10(x + y) + 4x - 2y = 20 \text{ (Since } x + 4y = 10)$$

$$10x + 10y + 4x - 2y = 20$$

$$14x + 8y = 20$$

Now $x + 4y = 10$

$14x + 8y = 20$ and $4 : 8 :: 10 : 20$

From the *Sūtra*, $x = 0$ and $4y = 10$, i.e., $8y = 20$ $y = 10/4 = 2\frac{1}{2}$

Thus $x = 0$ and **y =** 2.5 is the solution.

*****

8. *Sūtra 7:* संकलनव्यवकलनाभ्याम् *– Saṅkalana–Vyavakalanābhyām*

– By Addition and by Subtraction

Saṅkalana – Joining, adding, holding together; *Vyavakalana* – Separation, subtraction, deduction; and *Bhyām* – Using both.

The *Sūtra* reads: संकलनव्यवकलनाभ्याम् – *Saṅkalana–Vyavakalanābhyām* which translated into English, simply says "By Addition and by Subtraction". Its application and modus operandi are:

It can be applied in solving a special type of simultaneous equations where the coefficients of x and the coefficients of y, are found interchanged (of course the absolute values and not the actual value including sign).

Example 1:

$$45x - 23y = 113 \quad (1)$$
$$23x - 45y = 91 \quad (2)$$

In the conventional method we have to make either the coefficient of x or coefficient of y in both the equations as equal. For that we have to multiply equation (1) by 45 and equation (2) by 23 and subtract to get the value of x and then substitute the value of x in one of the equations to get the value of y. Or we have to multiply equation (1) by 23 and equation (2) by 45 and then subtract to get value of y and then substitute the value of y in one of the equations, to get the value of x. It is difficult process to think of.

Using *Saṅkalana Vyavakalanābhyām*

Add the equations,

$$\text{i.e., } (45x - 23y) + (23x - 45y) = 113 + 91$$
$$\text{i.e., } 68x - 68y = 204$$
$$\therefore \quad x - y = 3 \quad (3)$$

Subtract one from the other,

$$\text{i.e., } (45x - 23y) - (23x - 45y) = 113 - 91$$
$$\text{i.e., } 22x + 22y = 22$$
$$\therefore \quad x + y = 1 \quad (4)$$

and by adding (3) and (4), we get $x = 2$ and by substituting we get $y = -1$

Example 2:

$$1955x - 476y = 2482$$
$$476x - 1955y = -4913$$
$$\text{Add, } 2431(x - y) = -2431 \quad \therefore\ x - y = -1$$
$$\text{Subtract, } 1479(x + y) = 7395 \quad \therefore\ x + y = 5$$
$$\text{Once again add, } 2x = 4 \quad \therefore x = 2$$
$$\text{Subtract } -2y = -6 \quad \therefore y = 3$$

*****

9. *Sūtra* 8: पूरणापूरणाभ्याम् – *Pūraṇāpūraṇābhyām*

– By the Completion or Non–Completion

Pūraṇa – complete; *Apūraṇa* – incomplete; and *Bhyām* – Using both.

The *Sūtra* reads: पूरणापूरणाभ्याम् – *Pūraṇāpūraṇābhyām*, which translated into English, simply says "By the Completion and/ or Non–Completion". Its application and modus operandi are:

The *Sūtra* can be taken as ***Pūraṇa*** – ***Apūraṇābhyām*** which then means by the completion and/ or non–completion. ***Pūraṇa*** is well known in the present system. We can see its application in solving the roots for general form of quadratic equation.

We have: $ax^2 + bx + c = 0$

$x^2 + (b/a)x + c/a = 0$ (dividing by a)

$x^2 + (b/a)x = -c/a$

Completing the square (i.e. ***Pūraṇā***) on the L.H.S.

$$x^2 + (b/a)x + (b^2/4a^2) = -c/a + (b^2/4a^2)$$

$$[x + (b/2a)]^2 = (b^2 - 4ac) / 4a^2$$

Proceeding in this way we finally get $x = \dfrac{-b \pm \sqrt{b^2 - 4ac}}{2a}$

Now we apply ***Pūraṇa*** to solve problems.

Example 1. $x^3 + 6x^2 + 11x + 6 = 0.$

Since $(x + 2)^3 = x^3 + 6x^2 + 12x + 8$

Adding (x + 2) to both sides

We get $x^3 + 6x^2 + 11x + 6 + x + 2 = x + 2$

i.e. $x^3 + 6x^2 + 12x + 8 = x + 2$

i.e. $(x + 2)^3 = (x + 2)$

This is of the form $y^3 = y$ if y is taken to be = x + 2

Solution y = 0, 1, – 1

i.e. x + 2 = 0, 1, –1 and hence x = –2,–1,–3

Example 2: $x^3 + 8x^2 + 17x + 10 = 0$

we know $(x + 3)^3 = x^3 + 9x^2 + 27x + 27$

Hence adding the term $(x^2 + 10x + 17)$ on the both sides, we get

$$x^3 + 8x^2 + 17x + x^2 + 10x + 17 = x^2 + 10x + 17$$

i.e. $x^3 + 9x^2 + 27x + 27 = x^2 + 6x + 9 + 4x + 8$

i.e. $(x + 3)^3 = (x + 3)^2 + 4\,(x + 3) - 4$

$y^3 = y^2 + 4y - 4$ for $y = x + 3$

$y = 1, 2, -2.$

Hence $x = -2, -1, -5$

*****

10. *Sūtra* 9: चलनकलनाभ्याम् – *Calanakalanābhyām*

– Differential Calculus

Calana – Moving, movable, shaking, vibrating, any motion or movement; *Kalana* – causing, effecting, inciting; and *Bhyām* – Use both

The *Sūtra* reads: चलनकलनाभ्याम् – *Calana–kalanābhyām*, which translated into English, simply says *Differential Calculus.* Its application and modus operandi are:

In his book on *Vedic* Mathematics *Swāmiji* has mentioned this *Sūtra, Calana–kalanābhyām* only in two places. Its meaning can be taken as "Sequential motion".

i) In the first instance it is used to find the roots of a quadratic equation

$$7x^2 - 11x - 7 = 0.$$

Swāmiji called the *Sūtra* as calculus formula. Its application at that point is as follows.

Now by calculus formula we say: $14x - 11 = \pm\sqrt{317}$

ii) At the Second instance under the chapter "Factorisation and Differential Calculus" for factorising expressions of 3rd, 4th and 5th degree, the procedure is mentioned as "*Vedic Sūtra*–s relating to

Calana – *Kalana* – Differential Calculus".

(i) Solution of quadratic equation:

Example 1: $x^2 - 5x + 6 = 0$

Its binomial factors are $(x - 2)(x - 3) = 0$

Hence $x = 2, 3$

Using the *Sūtra*, the discriminant (D) of the equation is given by:

$b^2 - 4ac$, in this case, $D = +1, -1$

The first differential of the term is given as $2x - 5$

Now equating the differential to the discriminant

$$2x - 5 = +1, -1$$

$$x = 3 \text{ or } x = 2.$$

Example 2: $7x^2 - 5x - 2 = 0$

The discriminant is 81

The first differential is $14x - 5$

Now, $14x - 5 = +9, -9$

$x = 1$ or $-2/7$

Example 3: $x^2 - 12x - 1 = 0$

The discriminant is 148

Differential is $2x - 12$

$2x - 12 = (148)^{1/2}$ the value of x can be calculated

*****

11. *Sūtra* 10: यावदूनम् – *Yāvadūnam*

– By the Deficiency

Yāvad – as large as, as much as, as many, as frequent, as long as, as old as; and *Ūnam* – less.

The *Sūtra* reads: यावदूनम् – *Yāvadūnam*, which translated into English, simply says "By the Deficiency". Its application and modus operandi are:

(i) Square of numbers

In this *Sūtra* for any problem, we have to find the nearest base. The bases are 10, 100, 1000, etc.

Example 1: To calculate 12^2

12 = 10 + 2, using base 10

Left part is calculated by adding the number through which the given number is greater than the base i.e. 10.
Hence Left part = 12 + 2 = 14
The Right part is calculated by simply squaring the excess.
Hence Right part = $2^2 = 4$
The answer is 144.

Example 2: To calculate 14^2

Base is 10
Left part is 14 + 4 = 18
Right part is $4^2 = 16$
From 16, 6 is used and 1 is carried over to the Left part.
Hence, the answer is 18 / 16
18 + 1 / 6 = 19 / 6 = 196

Note – We can also convert the main base to a suitable base using working base to find the square of the number.

Example 3: To calculate 54^2

$54 = 50 + 4$

Working base is 50, can be converted as base = x = 10 / 50

the Left part is 54 + 4 = 58

Divide the Left part as obtained from old method and divide by suitable base, x.

58 / (1 / 5) = 58 x 5 = 290

The Right part is obtained as previously.

$4^2 = 16$

6 is used and 1 is carried over.

Hence the answer is 290 / 16 = 290 + 1 / 6 = 291 / 6 = 2916.

In all the above cases, the number whose square is to be calculated is greater than the base. For finding the square of the number less than the base, similar procedure is carried out.

Example 4: To calculate 8^2

$8 = 10 - 2$

Left part is calculated by subtracting the difference of the number from the base.

Left part = 8 – 2 = 6

The Right part is calculated as $2^2 = 4$

The answer is 64.

Example 5: To calculate 88^2

$88 = 90 - 2$

The suitable base is 10 / 90 = 1 / 9

Left part is 88 – 2 = 86 / (1/9) = 86 x 9 = 774

Right part is $2^2 = 4$

The answer is 774 / 4 = 7744

(ii) Cubes of numbers – The techniques for squaring numbers can also be applied to calculate higher powers too. Here, a technique for cubing of numbers is shown.

Example 6: To calculate 14^3

$$14 = 10 + 4$$

i) To get the Left part, add twice the excess or deficiency to the number.

$$14 + 8 = 22$$

ii) Now the middle part
The new excess is 22 – 10 = 12
Multiply the new excess by the initial excess
Middle part is 12 x 4 = 48

iii) The Right part is as follows
Cube the excess or deficiency

$$4^3 = 64$$

iv) Hence the answer is

22 / 48 / 64
22 / 48 + 6 / 4 = 22 / 54 / 4
22 + 5 / 4 / 4
27 / 4 / 4
2744

Example 7: To calculate 24^3:

Write the cube of the leftmost digit i.e., $2^3 = 8$

8

Multiply this result by the ratio between the units digit and the tens digit of the given number. Write this product by the side –8 x (4 / 2) = 16

8 16

Repeat the above process two more times. The result of the previous multiplication = 16. Hence, we get – 16 x (4 / 2) = 32

8 16 32

Now, the result of previous multiplication = 32.

Multiplying this result by the ratio, we get 32 x (4 / 2) = 64

8 16 32 64

Note that 64, which was obtained last, is the cube of the units' place –4 of the given number.
Now, take the middle two numbers (16 and 32). Double them and write them below the numbers written already, i.e., 16 x 2 = 32 and 32 x 2 = 64

8	16	32	64
		32	64

After this, just add up the numbers written. We get the cube of the number.

The addition is done as follows.

We have added up the numbers in each stage to a single digit, with the carry produced being taken to the next stage. For '64', we write the units place '4' as the digit in the answer and the rest '6' as the carry for the next stage. In the next stage, we add up '64', '32' and '6' (previous carry) to get '102' as the result. Of this '102', '2' is placed as a digit in the answer and the rest of the digits are taken as carry. The addition is continued further in the same way and we get '13824' as the cube of '24'.

Example 8: As another example, let us find 35^3. Here, the ratio of our multiplication is (5/3). So, proceeding in the usual way, we get

27	45	75	125

In the next step, we double 45 and 75.

27	45	75	125
	90	150	

And finally, we add up the numbers.

15	23	12	
27	45	75	125
	90	150	
42	8	7	5

Thus, $35^3 = 42875$.

Example 9: To calculate 106^3

$$106 = 100 + 6$$

i) To get the Left part, add twice the excess or deficiency to the number.

$$106 + 12 = 118$$

ii) Now the middle part

The new excess is 118 – 100 = 18

Multiply the new excess by the initial excess

$$18 \times 6 = 108$$

The middle portion of the product 1 is carried over, 08 in the middle.

I.e. 106^3 = 118 / 08 / – – – – –

1

iii) The last portion of the product is cube of the initial excess.

I.e. $6^3 = 216$.

16 in the last portion and 2 carried over.

iv) i.e. 106^3 = 118 / 081 / 16 = 1191016

2

Like many other places, here too, the use of *miśrank* (vinculum – an overview is given at the end of this *sūtra*) makes things easier.

Let us take an example. To find 99^3:

We know that $9^3 = 729$. The normal way would be

241	225	72	
729	729	729	729
	1458	1458	
970	2	9	9

However, if we use *miśrank*, the same problem can easily be solved as follows.

99 can be represented using *miśrank* as $10\bar{1}$

Here, we may consider 10 similar to the tens digit in the previous examples. Hence, the common ratio is (–1) / 10.

$$10^3 = 1000.$$
$$1000 \times (-1)/10 = -100.$$
$$-100 \times (-1) / 10 = 10$$
$$10 \times (-1)/10 = -1$$

Hence, proceeding as usual, we get

$$\begin{array}{cccc} 1000 & -100 & 10 & -1 \\ & -200 & 20 & \\ \hline & & & \overline{1} \\ \hline \end{array}$$

Here, we write –1 using *miśrank*. In the next step, we get

$$\begin{array}{cccc} & 3 & & \\ 1000 & -100 & 10 & -1 \\ & -200 & 20 & \\ \hline & & & \overline{1} \\ \hline \end{array}$$

i.e., 20 + 10 = 30 => carry = 3.
Again, continuing the addition, we get

$$\begin{array}{cccc} & 3 & & \\ 1000 & -100 & 10 & -1 \\ & -200 & 20 & \\ \hline & & 0 & \overline{1} \\ \hline \end{array}$$

i.e., –200 + –100 + 3 = –297. The unit digit 7 is written in *miśrank* form and –29 is taken as the negative carry. Thus, finally we get

$$\begin{array}{cccc} -29 & 3 & & \\ 1000 & -100 & 10 & -1 \\ & -200 & 20 & \\ \hline 971 & \overline{7} & 0 & \overline{1} \\ \hline \end{array}$$

Eliminating the negative digits from $971\overline{7}0\overline{1}$ we get 970299 as the result.

To more clearly understand this method let us consider

$$(a + b)^3 = a^3 + 3a^2b + 3ab^2 + b^3$$

This can be written as

$$a^3 + a^2b + ab^2 + b3 + 2a^2b + 2ab^2$$

i.e.

$$\begin{array}{llllllll} a^3 & + & a^2b & + & ab^2 & + & b^3 \\ & + & 2a^2b & + & 2ab^2 & & \\ \hline a3 & + & 3a^2b & + & 3ab^2 & + & b^3 \\ \hline \end{array}$$

It can be observed that a^3, a^2b, ab^2, b^3 are in Geometric progression with common ratio (b/a).

That is, $a^3 \times (b/a) = a^2b$

$$a^2b \times (b/a) = ab^2$$

$$ab^2 \times (b/a) = b^3$$

In our process of cubing, 'a' & 'b' are the tens and units digits respectively. To compute the terms of the Geometric series, we progressively multiply by (b/a). This gives

$$a^3 + a^2b + ab^2 + b^3$$

Then, we double the two middle terms and add them up. i.e., $(a^2b + ab^2) \times 2 = 2a^2b + 2ab^2$

$$\begin{array}{l} a^3 + a^2b + ab^2 + b^3 \\ \quad + 2a^2b + 2ab^2 \\ \hline a^3 + 3a^2b + 3ab^2 + b^3 \\ \hline \end{array}$$

This gives the required cube of the number i.e., $(a + b)^3$

An overview of **vinculum**:

A vinculum is a horizontal line used in mathematical notation for a specific purpose. It may be placed as an overline (or underline) over (or under) a mathematical expression to indicate that the expression is to be considered grouped together. Historically, vincula were extensively used to group items together, especially in written mathematics, but in modern mathematics this function has almost entirely been replaced by the use of parentheses. Today, however, the common usage of a vinculum is to indicate the repetend of a repeating decimal is a significant exception and reflects the original usage.

Vinculum is Latin for ‘bond’, ‘fetter’, ‘chain’, or ‘tie’, which is suggestive of some of the uses of the symbol

A vinculum can indicate a line segment where A and B are the endpoints - $\overline{AB}$

A vinculum can indicate the repetend of a repeating decimal value:

$1/7 = 0.\overline{142857} = 0.142857142857142857\ldots$

Similarly, it is used to show the repeating terms in a periodic continued fraction. Quadratic irrational numbers are the only numbers that have these.

Its main use was as a notation to indicate a group (a bracketing device serving the same function as parentheses): $(a - \overline{b + c})$, meaning to add b and c first and then subtract the result from a, which would be written more commonly today as (a – {b + c}). Parentheses, used for grouping, are only rarely found in the mathematical literature before the eighteenth century.

The vinculum is used as part of the notation of a radical to indicate the radicand whose root is being indicated. In the following, the quantity ab + 2 is the whole radicand, and thus has a vinculum over it $\sqrt[n]{ab + 2}$.

In 1637 Descartes was the first to unite the German radical sign √ with the vinculum to create the radical symbol in common use today.

The symbol used to indicate a vinculum need not be a line segment (overline or underline); sometimes braces can be used (pointing either up or down).

The vinculum number is written as below:

$$9 = 10 - 1 = 1\bar{1}$$
$$8 = 10 - 2 = 1\bar{2}$$
$$28 = 30 - 2 = 3\bar{2}$$
$$47 = 50 - 3 = 5\bar{3}$$
$$174 = 200 - 30 + 4 = 2\bar{3}4$$

A vinculum number is a takeaway or minus number. They are really efficient. Writing any number in this manner makes the operations easier. Any high value digit can easily be written as small value digit so that the operation becomes stress free. In other words only the digits 0 to 5 are used and the other digits 6 to 9 will be converted.

*****

12. *Sūtra* 11: व्यष्टिसमष्टिः – *Vyaṣṭisamasṭiḥ*

– Specific and General

Vyaṣṭi – singleness, individuality, a separated aggregate (such as man, viewed as a part of a whole (e.g. of the Universal Soul) while himself composed of individual parts, the state of individuality and totality regarding (a group of objects) singly or individually; and *Samaṣṭi* – collective existence, collectiveness, an aggregate, totality, regarding a group of objects collectively.

The *Sūtra* reads: व्यष्टिसमष्टिः – *Vyaṣṭisamasṭiḥ*, which translated into English, simply says "Specific and General". Its application and modus operandi are:

This ***Sūtra*** is also interpreted as "Whole as one and one as whole" and again as "Part and Whole".

Vyaṣṭisamasṭiḥ Sūtra teaches one how to use the average or exact middle binomial for breaking the biquadratic down into a simple quadratic by the easy device of mutual cancellations of the odd powers.

However the modus operandi is explained in detail alongwith corollary 11– लोपनस्थापनाभ्याम् – *Lopanasthāpanābhyām* – By Alternative Elimination and Retention.

Instead of explaining these 2 separately, it is a little easier to understand when explained in a combined manner and hence not discussed here separately.

*****

13. *Sūtra* 12: शेषाण्यङ्केन चरमेण – *Śeṣānyaṅkena Carameṇa* – The Remainders by the Last Digit

Śeṣa – remainder, what's left; *Anya* – other; *Ank* – connect, join; and *Carama* – the last, final, outermost.

The *Sūtra* reads: शेषाण्यङ्केन चरमेण – *Śeṣānyaṅkena Carameṇa*, which translated into English, simply says "The Remainders by the Last Digit". *Caramāṇka* means the last digit. Its application and modus operandi are:

Division by prime numbers

For using this method, first we have to compute the remainders and then multiply the remainders with the last digits and put the last digit of the multiplicand.

Example 1: Let us consider 1 ÷ 7

Conventional method:

```
7) 10 (0.142857
    7
   ---
    30
    28
   ---
     20
     14
   ---
      60
      56
   ---
       40
       35
   ---
       50
       49
   ---
        1
```

Vedic method:

When dividing 1(0) by 7 we get the remainder as 3. Therefore, dividing 3 by 7 will consequently lead to remainder 9 (= 3 x 3). But since 9 is more than 7 the remainder would be 2, hence the remainder sequence is: 3, 2

Now we divide 2 by 7 and we get the remainder as 6 (3 x 2), that is 3, 2, 6

On rearranging the remainders we get 3, 2, 6, 4, 5, 1

We stop when the remainder sequence starts to repeat itself. Now, we multiply these remainders by the last digit (7) of the denominator and keep only the first digit. Hence

7 x 3 = 21 put down 1
.1
7 x 2 = 14 => put down 4
.1 4
7 x 6 = 42 => put down 2
.1 4 2
on continuing we get
.1 4 2 8 5 7
Hence the answer is 1 ÷ 7 = 0.142857142857...

Example 2: 1 ÷ 13

On rearranging the remainders we get 10, 9, 12, 3, 4, 1

Now multiply these with 3, the last digit of the dividend 13:

10 x 3 = 30
9 x 3 = 27
12 x 3 = 36
3 x 3 = 9
4 x 3 = 12
1 x 3 = 3

We put the last digit of each product down in our answer.
Hence the answer is 0.076923

We know that 1 ÷ 7 = 0.142857
Hence 2 ÷ 7 = 2 x (1 ÷ 7) = 0.285714
3 ÷ 7 = 3 x (1 ÷ 7) = 0.428571

Similarly we calculate 2 ÷ 13, 3 ÷ 13, 4 ÷ 13. etc., till 12 ÷ 13 using the above method

Example 3: 1 ÷ 11

1 ÷ 11 can be written as 9 ÷ 99
Now we can calculate 1 ÷ 99 using the ***Ekādhikena*** *Sūtra* and then multiply the answer by 9.

Example 4: 1 ÷ 17

1 ÷ 17 can be written as 7 ÷ 119. The value of 1 ÷ 119 can be calculated by ***Ekādhikena*** *Sūtra*. We multiply by 7 to get the value of 1 ÷ 17

Example 5: 1 ÷ 23

1 ÷ 23 can be written as 3 ÷ 69.
Now again 1 ÷ 69 can be calculated by ***Ekādhikena*** *Sūtra*.
To get 1 ÷ 23 we multiply the answer with 3.

*****

14. *Sūtra* 13: सोपान्त्यद्वयमन्त्यम् – *Sopāntyadvayamantyam* – The Ultimate and Twice the Penultimate

Sa – with; *Upa* – near; *Antya* – the last, the one after; *Dvayam* – two, two-fold, dual, double, couple, pair; and *Antyam* – the last, the end, final thing, the one after. (As we get close there will be two conclusions).

The *Sūtra* reads: सोपान्त्यद्वयमन्त्यम् – *Sopāntyadvayamantyam*, which translated into English, simply says "The Ultimate and Twice the Penultimate". Its application and modus operandi are:

Solution of miscellaneous simple equations: The *Sūtra* is useful in finding the solution of equations of the type

$$\frac{1}{AB} + \frac{1}{AC} = \frac{1}{AD} + \frac{1}{BC}$$

According to the *Sūtra*, the last + twice the penultimate is 0.

Example 1:

$$\frac{1}{(x^2 + 7x + 12)} + \frac{1}{(x^2 + 8x + 15)} = \frac{1}{(x^2 + 9x + 18)} + \frac{1}{(x^2 + 9x + 20)}$$

First let us solve each quadratic mentally

$$\frac{1}{(x + 3)(x + 4)} + \frac{1}{(x + 3)(x + 5)} = \frac{1}{(x + 3)(x + 6)} + \frac{1}{(x + 4)(x + 5)}$$

Using the *Sūtra* we get (P is penultimate and L is last):

$$2P + L = 0$$
$$2(x + 5) + x + 6 = 0$$
$$x = -16 \div 3$$

Example 2:

$$\frac{1}{(2x+1)(3x+2)} + \frac{1}{(2x+1)(4x+3)} = \frac{1}{(2x+1)(5x+4)} + \frac{1}{(3x+24)(4x+3)}$$

$$2P + L = 0$$
$$2(4x+3) + (5x+4) = 0$$
$$x = -10/13$$

$$\frac{1}{(x+2)(x+3)} + \frac{1}{(x+2)(x+4)} = \frac{1}{(x+2)(x+5)} + \frac{1}{(x+3)(x+4)}$$

Here according to this *Sūtra* (L + 2P) (the last + twice the penultimate)

$$= (x+5) + 2(x+4) = 3x + 13 = 0$$

$\therefore$ $x = 4\ 1/3$

The proof follows:

$$\frac{1}{(x+2)(x+3)} + \frac{1}{(x+2)(x+4)} = \frac{1}{(x+2)(x+5)} + \frac{1}{(x+3)(x+4)}$$

$$\therefore \quad \frac{1}{(x+2)(x+3)} - \frac{1}{(x+2)(x+5)} = \frac{1}{(x+3)(x+4)} - \frac{1}{(x+2)(x+4)}$$

$$\therefore \quad \frac{1}{(x+2)}\left\{\frac{2}{(x+3)(x+5)}\right\} = \frac{1}{(x+4)}\left\{\frac{-1}{(x+2)(x+3)}\right\}$$

Removing the common factors (x + 2) and (x + 3)

$$\frac{2}{(x+5)} = \frac{-1}{(x+4)}$$

i.e. $$\frac{2}{L} = \frac{-1}{P}$$

$$\frac{1}{(x+2)(x+3)}+\frac{1}{(x+2)(x+4)}=\frac{1}{(x+2)(x+5)}+\frac{1}{(x+3)(x+4)}$$

$\therefore \quad L + 2P = 0$

The General Algebraic Proof is as follows:

$$\frac{1}{AB}+\frac{1}{AC}=\frac{1}{AD}+\frac{1}{BC}$$

(Where A, B, C and D are in Arithmetic Progression with d as common difference)

$$\frac{1}{A(A+d)}+\frac{1}{A(A+2d)}=\frac{1}{A(A+d)}+\frac{1}{(A+d)(A+2d)}$$

$$\therefore \quad \frac{1}{A(A+d)}-\frac{1}{A(A+3d)}=\frac{1}{(A+d)(A+2d)}+\frac{1}{A(A+2d)}$$

$$\therefore \quad \frac{1}{A}\left\{\frac{2d}{(A+d)(A+3d)}\right\}=\frac{1}{(A+2d)}\times\frac{(-d)}{A(A+d)}$$

Cancelling the common factors A(A + d) of the denominators and d of numerator:

$$\therefore \quad \frac{2}{(A+3d)}=\frac{-1}{(A+2d)}$$

In other words $\dfrac{2}{L}=\dfrac{-1}{P}$

$\therefore \quad L + 2P = 0$

*****

15. *Sūtra* 14: एकन्यूनेन पूर्वेण – *Ekanyūnena Pūrveṇa*

– By One Less than the One Before

Eka – One; *Anya* – other; *Ūnena* – Less; and *Pūrveṇa* – What used to be before.

The *Sūtra* reads: एकन्यूनेन पूर्वेण – *Ekanyūnena Pūrveṇa*, which translated into English, simply says "By One Less than the One Before". Its application and modus operandi are:

1) The use of this *Sūtra* in case of multiplication by 9,99,999 follows:

Method:

a) The left hand side digit(s) is obtained by applying the *Ekanyūnena Pūrveṇa* i.e. by deducting 1 from the left side digit(s) e.g. (1) 7 x 9; 7 – 1 = 6 (Left digit)
b) The right hand side digit is the complement or difference between the multiplier and the left hand side digit (digits).
i.e. 7 x 9 – Right digit is 9 – 6 = 3.
c) The product of the two numbers give the answer; i.e. 7 x 9 = 63.

Example 1: 8 x 9

Step (a) gives 8 – 1 = 7 (Left Digit)
Step (b) gives 10 – 8 = 2 (Right Digit)
Step (c) gives the answer 72

Example 2: 356 x 999

Step (a): 356 – 1 = 355
Step (b): 1000 – 356 = 644
Step (c): 356 & 999 = 355644

Example 3: 878 x 9999

Step (a): 878 – 1 = 877
Step (b): 10000 – 878 = 9122
Step (c): 877 & 9122 = 8779122

What happens when the multiplier has lesser digits?
i.e. for problems like 42 x 9, 124 x 9, 26325 x 99 etc.

For this let us have a re–look in to the process.

Multiplication table of 9:

	a	b
2 x 9 =	1	8
3 x 9 =	2	7
4 x 9 =	3	6
------—		
8 x 9 =	7	2
9 x 9 =	8	1
10 x 9 =	9	0

It can be observed that the left hand side of the answer is always one less than the multiplicand (here multiplier is 9) as read from Column (a) and the right hand side of the answer is the complement of the left hand side digit from 9 as read from Column (b)

Multiplication table when both multiplicand and multiplier are of 2 digits:

11 x 99 = 1089
12 x 99 = 1188
13 x 99 = 1287

---

18 x 99 = 1782
19 x 99 = 1881
20 x 99 = 1980

The rule mentioned in the case of above table also holds good here. Further we can state that the rule applies to all cases, where the multiplicand and the multiplier have the same number of digits.

Let us consider the following Tables.

(i)

	a	b
11 x 9 =	9	9
12 x 9 =	10	8
13 x 9 =	11	7
18 x 9 =	16	2
19 x 9 =	17	1
20 x 9 =	18	0

(ii)

21 x 9 =	18	9
22 x 9 =	19	8
23 x 9 =	20	7
28 x 9 =	25	2
29 x 9 =	26	1
30 x 9 =	27	0

(iii)

35 x 9 =	31	5
46 x 9 =	41	4
53 x 9 =	47	7
67 x 9 =	60	3

—————————so on.

From the above tables the following points can be observed:

1) Table (i) has the multiplicands with 1 as first digit except the last one. Here left part of products is uniformly 2 less than the multiplicands. Hence also with 20 x 9.

2) Table (ii) has the same pattern. Here left part of products is uniformly 3 less than the multiplicands.

3) Table (iii) is of mixed example and yet the same result i.e. if 3 is first digit of the multiplicand then left part of the product is 4 less than the multiplicand; if 4 is first digit of the multiplicand then, left part of the product is 5 less than the multiplicand and so on.

4) The right hand side of the product in all the tables and cases is obtained by subtracting the right part of the multiplicand by *Nikhilam*.

Keeping these points in view we solve the below problems:

Example 4: 42 x 9

i) Divide the multiplicand (42) of by a vertical line or by the sign: into a right hand portion consisting of as many digits as the multiplier.

i.e. 42 has to be written as 4 / 2 or 4 : 2

ii) Subtract from the multiplicand one more than the whole excess portion on the left. i.e. left portion of the multiplicand is 4. One more than it 4 + 1 = 5.

We have to subtract this from the multiplicand

i.e. write it as

```
4 :  2
  : –5
----------
3 :  7
```

This gives the left part of the product. This step can be interpreted as "take the *Ekanyūnena* and subtract from the previous" i.e. the excess portion on the left.

iii) Subtract the right part of the multiplicand by *Nikhilam* process.

i.e. Right part of the multiplicand is 2. Its *Nikhilam* is 8.

It gives the right part of the product

i.e. answer is 3 : 7 : 8 = 378.

Thus 42 x 9 can be represented as

```
4 : 2
  :–5 : 8
--------------
3 : 7 : 8 = 378.
```

Example 5: 124 x 9

124 x 9 is 12 : 4
−1 : 3 : 6

---

11 : 1 : 6 = 1116

The process can also be represented as
124 x 9 = {124 – (12 + 1)} : (10 – 4) = (124 – 13) : 6 = 1116

Example 6: 15639 x 99

Since the multiplier has 2 digits, the answer is
{15639 – (156 + 1)}: (100 – 39) = (15639 – 157) : 61 = 1548261

*****

16. *Sūtra* 15: गुणितसमुच्चयः – *Guṇitasamuccayaḥ*

– The Product of the Sums

Guṇita – property, functionality, basic substance; and *samuccayaḥ* – whole

The *Sūtra* reads: गुणितसमुच्चय – *Guṇitasamuccayaḥ*, which translated into English, simply says "The Product of the Sums". Its application and modus operandi are:

The *Sūtra* means the product of the sum equals to the sum of the product.

Let us take a concrete example and see how this method can be made use of.

Suppose we have to factorise $x^3 + 6x^2 + 11x + 6$ and by some method, we know $(x + 1)$ to be one of the factors.

We first use the *Upa–sūtra* viz. *Ādyamādyenāntyamantyena*[1] and thus mechanically put down 2 and 6 as the first and the last coefficients in the quotient;

i.e. the product of the remaining two binomial factors.
But we already know that the Sc of the given expression is 24.

As the Sc of $(x + 1) = 2$ we therefore know that the Sc of the quotient must be 12.

As the first and the last digits thereof are already known to be 1 and 6, their total is 7. And therefore the middle term must be $12 - 7 = 5$.

Hence, the quotient is $x^2 + 5x + 6$.

This is a very simple and easy but absolutely certain and effective process.

*****

[1] Corollary 3 discussed later

17. *Sūtra* 16: गुणकसमुच्चयः – *Guṇakasamuccayaḥ*

– All the Multipliers

Guṇaka – quality, multiplier; and *samuccayaḥ* – whole

The *Sūtra* reads: गुणकसमुच्चय – *Guṇakasamuccayaḥ*, which translated into English, simply says "All the Multipliers". Its application and modus operandi are:

The *Sūtra* means the factors of the sum is equal to the sum of the factors. It means the product of the sum of the coefficients in the factors is equal to the sum of the coefficients in the product.

In symbols we may put this *Sūtra* as:

Sc of the product = Product of the Sc (in factors).

For example:

$(x + 7)(x + 9) = x^2 + 16x + 63$ and we observe
$(1 + 7)(1 + 9) = 1 + 16 + 63 = 80.$

Similarly in the case of cubics, biquadratics etc., also the same rule holds good.

For example:

$$(x + 1)(x + 2)(x + 3) = x^3 + 6x^2 + 11x + 6$$
$$2 \times 3 \times 4 = 1 + 6 + 11 + 6$$
$$= 24.$$

Thus if and when some factors are known this rule helps us to fill in the gaps.

In several places in the use of *Sūtra–s* the corollaries are *upa–Sūtra–s* are used in the middle.

One such instance is the *upa–Sūtra–s* (corollary 11) *Lopanasthāpanābhyām* which is cited verbatim here. The *Lopanasthāpanābhyām* corollary however removes the whole difficulty and makes the factorisation of a quadratic of this type as easy and simple as that of the ordinary quadratic.

The procedure is as follows:

Suppose we have to factorise the following long quadratic.

$$2x^2 + 6y^2 + 6z^2 + 7xy + 11yz + 7zx$$

i. We first eliminate by putting $z = 0$ and retain only x and y and factorise the resulting ordinary quadratic in x and y with *Adyamādyenāntyamantyena* (corollary 3)

ii. We then similarly eliminate y and retain only x and z and factorise the simple quadratic in x and z.

iii. With these two sets of factors before us we fill in the gaps caused by our own deliberate elimination of z and y respectively.

And that gives us the real factors of the given long expression.

The procedure is an argumentative one and is as follows:

If $z = 0$ then the given expression is $2x^2 + 7xy + 6y^2 = (x + 2y)(2x + 3y)$.
Similarly if $y = 0$ then $2x^2 + 7xz + 3z^2 = (x + 3z)(2x + z)$.

Filling in the gaps which we ourselves have created by leaving out z and y,

we get $E = (x + 2y + 3z)(2x + 3y + z)$

This *Lopanasthāpanābhyām* method of alternate elimination and retention will be found highly useful in finding H.C.F., in solid Geometry and in co–ordinate Geometry of the straight line, the hyperbola, the conjugate hyperbola, the asymptotes etc.

In the current system of Mathematics we have two methods which are used for finding the H.C.F. of two or more given expressions.

The first is by means of factorisation which is not always easy and the second is by a process of continuous division like the method used in the G.C.M. chapter of arithmetic. The latter is a mechanical process and can therefore be applied in all cases. But it is rather too mechanical and consequently long and cumbersome.

The *Vedic* methods provide a third method which is applicable to all cases and is at the same time free from all disadvantages.

It is mainly an application of the Formula 3 and corollaries 3 and 11 viz. *Ūrdhva–tiryagbhyām*, *Ādyamādyenāntyamantyena* and *Lopanasthāpanābhyām*

The procedure adopted is one of alternate destruction of the highest and the lowest powers by a suitable multiplication of the coefficients and the addition or subtraction of the multiples.

A concrete example will elucidate the process.

Suppose we have to find the H.C.F of $x^2 + 7x + 6$ and $x^2 – 5x – 6$

i.e. $x^2 + 7x + 6 = (x + 1)(x + 6)$ and $x^2 – 5x – 6 = (x + 1)(x – 6)$. H.C.F. is $(x + 1)$.

i. This is the first method.

ii. The second method the G.C.M. one is well–known and need not described here.

iii. The third process of *Lopanasthāpanābhyām* i.e. of the elimination and retention or alternate destruction of the highest and the lowest powers is explained below.

Let E1 and E2 be the two expressions. Then for destroying the highest power we should subtract E2 from E1 and for destroying the lowest one we should add the two.

The chart is as follows:

Addition:

$$\begin{array}{l} x^2 - 5x - 6 \\ x^2 + 7x + 6 \\ \hline 2x^2 + 2x \\ 2x(x + 1) \end{array}$$

Subtraction:

$$\begin{array}{r} x^2 + 7x + 6 \\ x^2 - 5x - 6 \\ \hline 12x + 12 \\ 12(x + 1) \end{array}$$

We then remove the common factor if any from each and we find x + 1 staring us in the face i.e. x + 1 is the H.C.F.

*****

18. Corollary 1 – आनुरूप्येण – *Ānurūpyeṇa* – Proportionately

– conformity, suitableness

The *Upa–Sūtra* reads: आनुरूप्येण – *Ānurūpyeṇa*, which translated into English, simply says '*Proportionately*'. Its application and modus operandi are:

This *Upa–Sūtra* is highly useful to find products of two numbers when both of them are near the common bases i.e. powers of 10.

In actual application, it connotes that in all cases where there is a rational ratio–wise relationship; the ratio should be taken into account and should lead to a proportionate multiplication or division as the case may be.

Example 1: To calculate 46 x 43

As per the previous methods, if we select 100 as base we get

Method 1: Take the nearest higher multiple of 10. In this case it is 50.

Treat it as 100 ÷ 2 = 50. Now the steps are as follows:

i) Choose the working base near to the numbers under consideration. i.e. working base is 100 ÷ 2 = 50

ii) Write the numbers one below the other

i.e. 46
43
------

iii) Write the differences of the two numbers respectively from 50 against each number on right side

i.e. 46 –04
43 –07

Write cross–subtraction or cross–addition as the case may be under the line drawn.

46 –04
43 –07

(46 – 7) or /
(43 – 4)

= 39

Multiply the differences and write the product in the left side of the answer.

46 –04
43 –07

39 / –4 x –7
= 28

vi) Since base is 100 ÷ 2 = 50, 39 in the answer represents 39 x 50.

Hence divide 39 by 2 because 50 = 100 ÷ 2

Thus 39 ÷ 2 gives 19½ where 19 is quotient and 1 is remainder.

This 1 as Reminder gives one 50 making the Left part of the answer 28 + 50 = 78 (or Remainder ½ x 100 + 28)

i.e. Right part 19 and Left part 78 together give the answer 1978

Method 2:

We take the same working base 50. We treat it as 50 = 5 x 10.

i.e. we operate with 10 but not with 100 as in previous method

```
        46                    –04
        43                    –07
   ______________________________
        (46 – 7) or  /        –4 x –7
        (43 – 4)    /
         = 39      /          = 28
   ______________________________
      39 x 5             8    /
       +2                    /
      (carried over)        /
          (195 + 2) ÷ 8 = 1978
```

Method 3: We take the nearest lower multiple of 10 since the numbers are 46 and 43 as in the first example, we consider 40 as working base and treat it as 4 x 10.

Let us see the all the three methods for a problem at a glance

Example 2: 492 x 404

Method 1: Working base = 1000 ÷ 2 = 500

```
     492   –008
     404   –096
    -----------
  2) 396  /  768     since 1000 is in operation
    -----------
     198  /  768     =  198768
```

Method 2: Working base = 5 x 100 = 500

```
        492                –008
        404                –096
   ________________________
        396      /    768
        x 5     /
   ________________________
     1980 / 768 = 198768
```

Method 3:

Since 400 can also be taken as working base, treat 400 = 4 x 100 as working base.

Thus

```
    492      092
    404      004 since 100 is in operation
  ---------------
    496   /  368
    x 4  /
  ---------------
  1984 / 368 = 198768
```

*****

19. Corollary 2 – शिष्यते शेषसंज्ञ: – *Śiṣyate Śeṣasamjñaḥ*

– The Remainder Remains Constant

The *Upa–Sūtra* reads: शिष्यते शेषसंज्ञ: – *Śiṣyate Śeṣasamjñaḥ*, which translated into English, simply says "The Remainder Remains Constant". Its application and modus operandi are:

The source concept here is the concept of equality of units, the pair of halves etc.

Format:

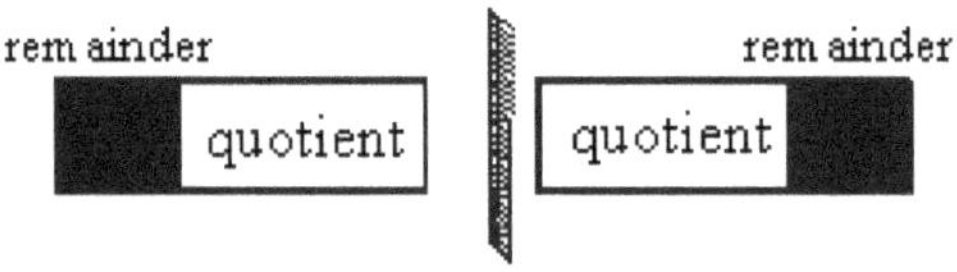

Basics/ technical words:

The basic/ technical word availed by the text of *Upa–Sūtra* 2 is '*Śeṣa*' (the remainder). The text as such is the definition of remainder:

Remainder is the one, which remains (as uncovered part of the divisor, on division).

The basic operation is the operation of 'division'. This corollary along with corollary 1 makes the Mathematics of corollaries a very interesting branch in itself.

The 'one' which permitted itself as a pair of halves, sequentially by the rule of symmetry of corollary 1 and the rule of equality of remainders of division of equals takes the process of partitioning of 'one' into as many equal parts as one would like to have.

This makes the external expansion of counts from one to two, two to three and so on parallel to the internal partitioning from wholesome one to a pair of parts, then to three parts, four parts and so on.

With this, the arrays of Mathematical tools become sufficiently large to work out very rich structures only in terms of 'parts', of which the working with 'half' is most rewarding.

This corollary is applicable only to a special case under the first corollary i.e. the squaring of numbers ending in 5 and other cognate numbers. Its wording is exactly the same as that of the *Sūtra* 2, which could be used at the outset for the conversion of 'vulgar' fractions into their recurring decimal equivalents. The Corollary now takes a totally different meaning and in fact relates to a wholly different setup and context.

Its literal meaning is the same as before (i.e. by one more than the previous one) but it now relates to the squaring of numbers ending in 5.

For example we want to multiply 15. Here the last digit is 5; and the "previous" one is 1.

Hence one more than that is 2.

Now Corollary in this context tells us to multiply the previous digit by one more than itself

I.e. by 2.

Hence the left hand side digit is 1 × 2 and the right hand side is the vertical multiplication product i.e. 25 as usual.

$$\frac{1 / 5}{2 / 25}$$

Thus $15^2 = 1 \times 2 / 25 = 2 / 25 = 225$;
Similarly $25^2 = 2 \times 3 / 25 = 6 / 25 = 625$;
$35^2 = 3 \times 4 / 25 = 6 / 25 = 1225$;

And so on.

*****

20. Corollary 3 – आद्यमाद्येन अन्त्यमन्त्येन – *Ādyamādyena Antyamantyena*

– The First by the First and the Last by the Last

Ādya – beginning*; mādye* – first*; nāntya* – last*; mantye* – last.

The *Upa–Sūtra* reads: आद्यमाद्येन अन्त्यमन्त्येन – *Ādyamādyena Antyamantyena*, which translated into English, simply says "The First by the First and the Last by the Last". In some schools this is mentioned as मध्यमध्येन आद्यमन्त्येन – *Madhyamadhyena Ādyamantyena*, which means "the product of the means equal to the product of the extremes". Its application and modus operandi are:

Example 1: Suppose we have to find out the area of a rectangular card board whose length and breadth are respectively 6 feet 4 inches and 5 feet 8 inches. Generally we proceed as:

Area = Length x Breadth

= 6′ 4″ x 5′ 8″

Since 1′ = 12″,= (6 x 12 + 4) x (5 x 12 + 8) conversion into single unit

= 76″ x 68″ = 5168 Sq. inches.

Since 1 sq. ft. = 12 x 12 = 144 sq. inches we have the area

```
5168   =   144) 5168 (35
----
144            432
               ----
                848
                720   i.e., 35 Sq. ft. 128 Sq. inches
               -----
                128
```

By *Vedic* principles we proceed in the way "the first by first and the last by last"

I.e. 6′ 4″ can be treated as 6x + 4 and 5′ 8″ as 5x + 8, where x = 1 feet = 12 in; x^2 is sq. ft.

Now (6x + 4) (5x + 8)

$= 30x^2 + (6 \times 8)x + (4 \times 5)x + 32$
$= 30x^2 + 68x + 32$
$= 30x^2 + (5x + 8)\,x + 32$ writing 68 as 5 x 12 + 8
$= 35x^2 + 8x + 32$
= 35 Sq. ft. + 8 x 12 Sq. in + 32 Sq. in
= 35 Sq. ft. + 96 Sq. in + 32 Sq. in
= 35 Sq. ft. + 128 Sq. in

It is interesting to note that a Mathematically untrained and even uneducated carpenter simply works in this way by mental argumentation. It goes in his mind like this:

6′ 4″
5′ 8″

First by first i.e. 6′ x 5′ = 30 sq. ft.

Last by last i.e. 4″ x 8″ = 32 sq. in.
Now cross wise 6 x 8 + 5 x 4 = 48 +20 = 68.
Adjust as many '12's as possible towards left as 'units'.

i.e. 68 = 5 x 12 + 8, 5 '12's as 5 square feet
Hence make the first 30 + 5 = 35 sq. ft.;

The remainder 8 becomes 8 x 12 square inches and going towards right i.e. 8 x 12 = 96 sq. in. gives 96 + 32 = 128 sq. in.

Thus the area is got in some sort of 35 sq. units and another sort of 128 sq. units.

i.e. 35 sq. ft. 128 sq. in.

Example 2:

4′ 6″
3′ 4″

4 x 3 / 4 x 4 + 6 x 3 / 6 x 4
= 16 + 18
= 12 / = 34 / = 24

= 2 x 12 + 10

Now 12 + 2 = 14, 10 x 12 + 24 = 120 + 24 = 144

Thus 4′ 6″ x 3′ 4″ = 14 Sq. ft. 144 Sq. inches.

Since 144 sq. in = 12 x 12 = 1 sq. ft. The answer is 15 sq. ft.

We can extend the same principle to find volumes of parallelepiped also.

Factorisation of quadratics:

The usual procedure of factorising a quadratic is as follows:

$$\begin{aligned} & 3x^2 + 8x + 4 \\ &= 3x^2 + 6x + 2x + 4 \\ &= 3x(x + 2) + 2(x + 2) \\ &= (x + 2)(3x + 2) \end{aligned}$$

But by mental process, we can get the result immediately. The steps follow:

Split the middle coefficient in to two such parts that the ratio of the first coefficient to the first part is the same as the ratio of the second part to the last coefficient.

Thus we split the coefficient of middle term of $3x^2 + 8x + 4$

i.e. 8 into two such parts 6 and 2 such that the ratio of the first coefficient to the first part of the middle coefficient

i.e. 3 : 6 and the ratio of the second part to the last coefficient, i.e. 2 : 4 are the same.

It is clear that 3 : 6 = 2 : 4.

Hence such a split is valid. Now the ratio 3 : 6 = 2 : 4 = 1 : 2 gives one factor x + 2.

ii) Second factor is obtained by dividing the first coefficient of the quadratic by the fist coefficient of the factor already found and the last coefficient of the quadratic by the last coefficient of the factor.

i.e. the second factor is

$$\frac{3x^2}{x} + \frac{4}{2} = 3x + 2$$

Hence $3x^2 + 8x + 4 = (x + 2)(3x + 2)$

Example 3: $4x^2 + 12x + 5$

i) Split 12 into 2 and 10 so that as per rule 4 : 2 = 10 : 5 = 2 : 1 i.e., 2x + 1 is first factor.

ii) Now

$$\frac{4x^2}{2x} + \frac{5}{1} = 2x + 5 \text{ is the second factor.}$$

It is evident that we have applied two *Upa–Sūtra–s*, ***Ānurūpyeṇa*** i.e. 'proportionality' and ***Ādyamādyenāntyamantyena*** i.e. "the first by the first and the last by the last" to obtain the above results.

Note: In store keeping, raw materials Management of accountancy we have a concept called FIFO (First in First Out). This formula justifiably fits in this accountancy usage also.

*****

21. Corollary 4 – केवलैः सप्तकं गुण्यात् – *Kevalaiḥ Saptakam Guṇyāt*

– For 7 the Multiplicand is 143

Saptaka – seven; *Kevalaiḥ* – 143 (using *Kaṭapayādi* method without transposing); and *Guṇyāt* – multiplicand;

The *Upa–Sūtra* reads: केवलैः सप्तकं गुण्यात् – *Kevalaiḥ Saptakam Guṇyāt*, which translated into English, simply says "For 7 the Multiplicand is 143". Its application and modus operandi are:

Hence this is interpreted as – in the case of seven, the multiplicand should be 143.

If we use the *Ekanyūnena* formula on multiplication:

In the case of 7 as denominator, 143 x 999 = 142/ 857.

There are six recurring decimal digits in the answer.

It can be observed that the rule of complements from 9 at work.

I.e. The corollary merely gives us the necessary clue to the first half of the decimal and also a simple device *Ekanyūnena* for arriving at the whole answer.

All this is achieved with the help of the easy alphabet code.

The result may therefore be formulated as:

$$\frac{1}{7} = \frac{143 \times 999}{999999} = 0.142857;$$

Let us consider some fractions with non–unity as numerator. The below chart may be observed:

$$\frac{1}{7} = 0.142857;$$

$$\frac{2}{7} = 0.285714;$$

$$\frac{3}{7} = 0.428571;$$

$$\frac{4}{7} = 0.571428;$$

$$\frac{5}{7} = 0.714285; \text{ and}$$

$$\frac{6}{7} = 0.857142;$$

Uniformity can be observed in the above chart – in all the 'Proper' fractions having 7 as their denominator:

- The same six digits are found
- They come up in the same sequence and in the same direction
- They however, start from a different starting point but travel in a 'cyclic' order
- With the aid of these rules, we can very easily obtain the recurring decimal equivalent of a vulgar fraction whose numerator is higher than 1.

*****

22. Corollary 5 – वेष्टनम् – *Veṣṭanam*

– By Osculation

Veṣṭanam – The act of surrounding, covering, encompassing or enclosing, wind or twist around.

The *Upa–Sūtra* reads: वेष्टनम् – *Veṣṭanam*, which translated into English, simply says "By Osculation". Its application and modus operandi are:

The Mathematical meaning of 'Osculation' is – the state or condition of touching or of being in immediate proximity or to touch at a point of common tangency to a line passing between two branches of a curve, each branch continuing in both directions of the line.

Though this corollary is given under *Sūtra* 5, शून्यं साम्यसमुच्चये *Śūnyam Sāmyasamuccaye* –"If the *Samuccaye* is the Same it is Zero", this works more in tandem with the first *Sūtra* एकाधिकेन पूर्वेण *Ekādhikena Pūrveṇa* – "By one more than the previous one".

Can 11 apples be divided equally among 4 people without cutting them?

A quick look will reveal that this cannot be done. This is because, 11 leaves a remainder of three when divided by 4. i.e. 11 is not **divisible** by 4.

A general class of such problems leads to an interesting branch of Mathematics dealing with divisibility. Divisibility can be determined through conventional techniques, only through actual cumbersome division. Shortcuts do exist, but they are only for a few numbers like 2, 3, 5, 9 and 11. *Vedic* techniques, on the other hand, provide us with a wealth of shortcuts, to determine divisibility of any number by any other number. These techniques can be applied mentally with great speed.

Simple Techniques:

Vedic techniques talk about *Veṣṭanas* or Osculators, which are to be used to determine divisibility. These *Veṣṭanas* can be thought about as a 'seed' using which divisibility can be determined. One member of this group of Osculators is the *Ekādhika* which literally translates to "One more".

General Illustration:

Example 1: Determine if 84 is divisible by 7.
To do this, first the *Ekādhika* of 7 has to be determined.

Square 7 to get 49.
The number "One more" than 4 is 5.
Hence the *Ekādhika* of 7 is 5.

The procedure is:–

Multiply the unit's digit of $8\underline{4}$ by 5:	4 x 5 = 20
Add the ten's digit of $\underline{8}4$ to the result:	8 + 20 = 28

The resulting number 28 is divisible by 7. Hence 84 is divisible by 7.

What about numbers with three or more digits?

The procedure is simply extended.

Example 2:

Determine if 112 is divisible by 7:

The *Ekādhika* of 7 as determined earlier is 5.

Multiply the unit's digit of $11\underline{2}$ by 5:	5 x 2 = 10
Add the tens digit of $1\underline{1}2$ to the result:	1 + 10 = 11
Now multiply 11 by 5:	11 x 5 = 55
Add the Hundreds Digit of $\underline{1}12$ to 55:	1 + 55 = 56

Now 56 is divisible by 7 and hence 112 is divisible by 7.

Example 3:

Let us, for instance, start with the number 7.
The *Ekādhika* for 7 is derived from 7 x 7 = 49 and is therefore 5.

The *Ekādhika* is the clinching test for divisibility; and the process by which it serves this purpose is technically called *Veṣṭana* or Osculation.

Suppose we do not know and have to determine whether 21 is divisible by 7.
We multiply the last digit, i.e. 1 by the *Ekādhika* (or positive oscillator, i.e. 5)

and

Add the product, i.e. 5 to the previous digit, i.e. 2 and thus get 7.
This process is called Osculation.

If the result of the osculation is the divisor itself or a repetition of a previous result, we can conclude that the given original dividend 21 is divisible by 7

Example 4:

Let us, now try with the number 13. The *Ekādhika* is 4.

Hence we go on multiplying leftward by 4. Thus,

13; 3 x 4 + 1 = 13
26; 6 x 4 + 2 = 26

and so on.

More examples:

1) 7 continually osculated by 5 gives 35, 28, 42, 14, 21 and 7.
2) 5 so osculated by 7 gives 35, 38, 59, 68, 62, 20, 2
3) 9 (by 7) gives 63, 27, 51, 12, 15, ...

The osculation of any number by its own *Ekādhika* will, as in the case of 7 or 13, go on be giving that very number or a multiple thereof. Thus:

23 osculated by 7 (its *Ekādhika*) gives 7 x 3 + 2 = 23;
46 osculated by 7 (its *Ekādhika*) gives 7 x 6 + 2 = 46;
and so on.

Whenever a problem of division comes up, we can adopt the following procedure.

Suppose we want to know whether 2774 is divisible by 19 or not (not by actual division):

Step 1: Find the *Ekādhika* Osculator – it is 2 here
Step 2: Multiply the last digit 4 by 2 = 8
Step 3: Add the product to the previous digit 7 = 15
Step 4: Multiply 15 by 2 = 30 – cast out 19 from 37 and put down the remainder 18 beneath 7
Step 5: Osculate that 18 with 2 to the left on the upper row and get 38;

2	7	7	4
		15	
	37		
	——		
	18		

2	7	7	4
19	18	15	

Since 19 is divisible by 19 the answer is 'Yes'.

<u>Rules for Determining *Ekādhikas*:</u>

Given below are the rules to determine the *Ekādhikas* of various numbers.
For convenience let the *Ekādhikas* be denoted as 'P'

For numbers ending with 9: 9, 19, 29... the *Ekādhika* is determined by adding 1 to the numbers and then dividing by 10.
e.g. P(29) = 29 + 1 = 30. 30 ÷ 10 = 3.

For numbers ending with 3: 3, 13, .. Multiply by 3 to get numbers ending with 9. Then follow the above rule to get the *Ekādhikas.*

i.e. P(13) = 13 x 3 = 39.
39 + 1 = 40.
40 ÷ 10 = 4.
P(13) = 4.

For numbers ending with 7: 7, 17, 27....Multiply by 7 to get numbers ending with 9. Then follow the first rule.

(e.g.) *Ekādhika* of 17 is – 17 x 7=119. 119 + 1 = 120. 120 ÷ 10 = 12.
P(17) = 12.

For Numbers Ending with 1: 1, 11, 21....Multiply with 9 to get numbers ending with 9. Then the usual rule is followed.

A simplification

Consider the problem – Is 4379 divisible by 29?
The Standard procedure would be:

1. Determine the *Ekādhika* of 29: 29 + 1 = 30. 30 ÷ 10 = 3. P = 3
2. Multiply the Unit's digit of 4379 by 3: 9 x 3 = 27
3. Add ten's digit of 4379 to 27: 27 + 7 = 34

At this stage the standard procedure would be to multiply 34 by 3 and add 3. However this results in a very cumbersome process. To avoid this, take the remainder of 34 when divided by 29 (the original number) to get 5. The procedure is carried out with 5 instead.

1. Multiply 5 by 3 to obtain 15: 5 x 3 = 15
2. Add Hundred's Digit of 4379 to 15: 15 + 3 = 18
3. Multiply 18 by 3 to get 54: 18 x 3 = 54
4. Add Thousand's digit of 4379 to 54 4 + 54 = 58

58 is divisible by 29. Hence 4379 is divisible by 29.

The simplified method: At any point of time during the application of the procedure, for determining the divisibility of a number **N** by a number **M** if the intermediate result becomes cumbersome, the remainder of the result when divided by M can be used instead of the result itself.

This technique is simple as long as the *Ekādhikas* are sufficiently small.

But what happens when they become huge?

The Principle of Negative Osculation: Just as *Ekādhika* is used for testing for divisibility, a negative osculator can be used in place of *Ekādhika*, to test for divisibility when the *Ekādhika* becomes huge. To avoid confusion, let this 'Negative' *Ekādhika* be denoted by 'Q'.

The Rules are:

For Numbers ending with 1: Simply drop the 1 in the unit's place.

For Other Numbers: Multiply with appropriate numbers to get a '1' in the unit's digit and then follow the above rule. That is:

For Numbers ending with 7: Multiply with 3, drop the 1 in the unit's place.

Example: Q(17) = 17 x 3 = 51, dropping 1 we get 5.

For Numbers ending with 3: Multiply with 7, drop the 1 in the unit's place

For Numbers ending with 9: Multiply with 9, drop the 1 in the unit's place.

Example: Q(19) = 19 x 9 = 171. Dropping the 1 in unit's place, we get 17.

A very interesting feature is that, "P and Q add up to the divisor".

Example: P(19) = 2, Q(19) = 17.
2 + 17 = 19.
This feature can be used to determine P from Q or vice versa.

Another feature is that, if the last digit of a number is 3, its $P < Q$.
If it is 7, $Q < P$.

While using the Negative Osculator Q, the procedure is altered. Instead of taking the Dividend as such, it is alternately marked as positive and negative. That is, unit place is marked as positive, tens as negative, hundreds as positive thousands as negative... and so on.

Example: Is 11234 divisible by 41?

The Negative Osculator for 41 is 4, obtained after dropping the 1.
11234 is not taken as such.

Instead, it is interpreted as +1 −1 +2 −3 +4.

Unit's place is positive and then the sign alternates.
The rest of the procedure remains same.

1) Multiply Unit's place of 11234 by 4: 4 x 4 = 16
2) Add −3 to 16: 6 + (−3) = 13
3) Multiply 13 with 4: 13 x 4 = 52 (52 can be simplified as 11)
4) Add 2 to 11 to get 13.
4) Multiply 13 with 4 to get 52, Simplify it as 11, add −1 to 11 to get 10.
5) Multiply 10 and 4 to get 40, add 1 to get 41.
41 is divisible by 41. Hence 11234 is divisible by 41.

Although we have solved the problem of getting struck up with huge *Ekādhikas* by introducing the concept of negative *Ekādhika*, cases where huge divisors are present still pose a problem. They still yield huge positive as well as negative *Ekādhikas*.

Bigger Divisors – Better Techniques

The cases encountered so far dealt only with small divisors and hence small osculators. However there are techniques to deal with bigger numbers. The bigger divisors can be grouped under 2 categories.

1) Divisors having a regularity:

Those which have a 9 or a series of 9's as their last digit. The *Ekādhika* of these numbers can be determined by dropping the series of 9's and adding a 1 to the remaining number. (e.g.) *Ekādhika* of 59999 is 6.

Numbers, which have 1 as their last digit preceded by a series of zeros – the Negative *Ekādhika* of these numbers can be determined by dropping all the zeroes followed by the ending 1. For instance – Negative *Ekādhika* of 50001 is 5.

Illustration:

Is 69492392 divisible by 199?

Here we can drop two 9's. It can be noted that for the 9's to be dropped, they should occur in the *last* few digits of the number and they should be **continuous**. The *Ekādhika* is 2, obtained by dropping the 9's and adding 1 to the result. Since two 9's were dropped, this number is denoted by P_2. The number 69492392 is not taken as such. Instead it is grouped into numbers of two digits each (since two 9's were dropped) and used.

(i.e.) 69492392 becomes 69 49 23 92. Now the usual method is carried out.

Multiply 92 by 2, add 23 to get 207. Simplify it to 8
Multiply 8 by 2, add 49 to get 65.
Multiply 65 by 2 add 69 to get 199.

199 divisible by 199. Hence 69492392 is divisible by 199.

2) Divisors that do not fall under the above category:

For divisors that do not fall into the above category, they are multiplied with suitable numbers to obtain numbers that end with a series of 9's or ending with a series of zeros followed by 1.

Example: What is the osculator for 857?

It can be observed that 857 x 7 = 5999.
Ekādhika of 5999 is 6.
6 is used for determining divisibility.

But there is one important point. It can be observed that we are actually testing the divisibility by 5999 instead of 857. So if the numerator is not divisible by 7, even if it is divisible by 857, the end result will not be divisible by 5999. This can be adjusted by testing the end number for divisibility by 857 instead of 5999.

*****

23. Corollary 6 – यावदूनं तावदूनं – *Yāvadūnam Tāvadūnam* – Lessen by the Deficiency

Yāvadūnam – How much ever is less; and *Tāvadūnam* – that much is less.

The *Upa–Sūtra* reads: यावदूनं तावदूनं – *Yāvadūnam Tāvadūnam*, which translated into English, simply says "Lessen by the Deficiency". Its application and modus operandi are:

(i) Squaring of numbers

Let us consider squaring of numbers close to the bases like 10, 100, 1000 etc.

Example 1: To calculate 97^2

$97 = 100 - 3$

The Left part of the answer is obtained by subtracting or adding the deficiency or excess of the number from the base.

Hence Left part of the answer is $97 - 3 = 94$

Right part of the answer is obtained by squaring the deficiency or excess of the number from the base. Hence the Right part of the answer is $3^2 = 9$.
Thus we obtain 94 / 09
9 is taken as 09, since the base is 100
The answer is 9409.

Example 2: To calculate 993^2

$993 = 1000 - 7$
Left part of the answer is $993 - 7 = 986$
Right part of the answer is $7^2 = 49$
986 / 049 (since base is 1000)
The answer is 986049

Example 3: To calculate 10025^2

$10025 = 10000 + 25$

Left part of the answer is $10025 + 25 = 10050$

Right part of the answer is $25^2 = 625$

Thus 10050 / 0625 (since base is 10000)

the answer is 100500625

If the number is not close to the base, the suitable base is used.

Example 4: To calculate 490^2

The base used is $500 = 5 \times 100$

$490 = 500 - 10$

Left part of the answer is calculated in the similar manner as in above problems.

But since the base is 500, the Left is multiplied to 5 – 5 times 100.

Left part of the answer is $490 - 10 = 480$

$480 \times 5 = 2400$

Right part of the answer is $10^2 = 100$

Thus, 240 / 100 (since the base is 500 – a multiple of 100)

The answer is 240100.

*****

24. Corollary 7 – यावदूनं तावदूनीकृत्यवर्गंच योजयेत् – *Yāvadūnam Tāvadūníkṛtya Vargañca Yojayet*

– Whatever the Deficiency lessen by that amount and set up the Square of the Deficiency

Yāvadūnam – Whatever is less; *Tāvadūníkṛtya* – reduce further; and *Vargañca Yojayet* – use this to sort or arrange in groups.

The *Upa–Sūtra* reads: यावदूनं तावदूनीकृत्य वर्गंच योजयेत् – *Yāvadūnam Tāvadūníkṛtya Vargañca Yojayet*, which translated into English, simply says "Whatever the Deficiency lessen by that amount and set up the Square of the Deficiency". Its application and modus operandi are:

This corollary arising out of the *Nikhilam Sūtra* is "whatever the extent of deficiency; lessen it still further to that very extent; and also set up square of that deficiency".

This corollary also means whatever the extent of its surplus, increment it still further to that very extent; and set up the square of that surplus.

This evidently deals with the squaring of numbers. This can be applicable to obtain squares of numbers close to bases of powers of 10.

Method 1: Numbers near and less than the bases of powers of 10.

Example 1: To calculate 9988^2. Base is 10,000.

Deficit = 10000 – 9988 = 12.
Square of deficit – $12^2 = 144$.
Deficiency subtracted from number – 9988 – 12 = 9976.
Answer is 9976 / 0144 (since the base is 10,000).

Example 2: To calculate 88^2. Base is 100.
Deficit is 100 – 88 = 12.
Square of deficit – $12^2 = 144$.
Deficiency subtracted from number – 88 – 12 = 76.
Now answer is 76 / $_1$44 = 7744 (since the base is 100)

Method 2: Numbers near and greater than the bases of powers of 10.

Example 3: To calculate 13^2

Base is 10, surplus is 3.
Surplus added to the number – 13 + 3 = 16.
Square of surplus – $3^2 = 9$
Answer is 16 / 9 = 169.

Example 4: To calculate 112^2

Base = 100, Surplus = 12,
Square of surplus – $12^2 = 144$
add surplus to number – 112 + 12 = 124.
Answer is 124 / $_1$44 = 12544

Example 5: To calculate 10025^2
= (10025 + 25) / 25^2
= 10050 / 0625 (since the base is 10,000)
= 100500625.

Method 3: This is applicable to numbers which are near to multiples of 10, 100, 1000, etc. For this we combine the corollaries, *Ānurūpyeṇa* and *Yāvadūnam Tāvadūníkṛtya Vargañca Yojayet* together.

Example 6: To calculate 388^2. The nearest base is 400.

We treat 400 as 4 x 100. As the number is less than the base we proceed as follows:
Deficit – 400 – 388 = 12
Since it is less than base, deduct the deficit i.e. 388 – 12 = 376.
Multiply this result by 4 (since the base is 4 times 100) – 376 x 4 = 1504
Square of deficit – $12^2 = 144$.
Hence answer is 1504 / $_1$44 = 150544 (since we have taken multiples of 100).

Example 7: To calculate 485^2. The nearest base = 500.

Treat 500 as 5 x 100 and proceed

485^2 = (485 – 15) / 15^2 (since the deficit is 15)

= 470 / $_2$25 since 500 base is taken as 5 x 100

x 5 and 2 of 225 is carried over

= 2350/ $_2$25

= 235225

Example 8: To calculate 5012^2. The nearest lower base is 5000 = 5 x 1000

Surplus = 12

5012^2 = (5012 + 12)/ 12^2

= (5024) / 144

x 5

= 25120/ 144

= 25120144

Cubing of Numbers:

Example 9: To find 106^3.

We proceed as follows:

i) For 106, Base is 100. The surplus is 6.
Here we add double the surplus i.e. 106 + 12 = 118.
This makes the left most part of the answer.
i.e. answer proceeds like 118 / – – – – –

ii) Put down the new surplus i.e. 118 – 100 = 18 multiplied by the initial surplus
i.e. 18 x 6 = 108.
Since base is 100, we write 108 in carried over form 108 i.e. $_1$08
As this is middle portion of the answer, the answer proceeds like 118 / $_1$08 /....

iii) Write down the cube of initial surplus i.e. $6^3 = 216$ as the last portion

i.e. right hand side last portion of the answer.

Since base is 100, write 216 as $_{2}16$ as 2 is to be carried over.

Answer is 118 / $_{1}08$ / $_{2}16$

Now proceeding from right to left and adjusting the carried over, we get the answer as

119 / 10 / 16 = 1191016.

*****

25. Corollary 8 – अन्त्ययोर्दशकेऽपि – *Antyayordaśake'pi*

– Last Totaling 10

Antya – last; *daśa* – ten; *yor* – digit; and *kepi* – total

The *Upa–Sūtra* reads: अन्त्ययोर्दशकेऽपि – *Antyayordaśake'pi*, which translated into English, simply says "Last Totaling 10". Its application and modus operandi are:

The Corollary signifies numbers of which the last digits added up give 10.

i.e. the Corollary is used in multiplication of numbers like:

25 x 25, 47 x 43, 62 x 68, 116 x 114.

It can be noted that in each case the sum of the last digits of the multiplier and the multiplicand is 10.

Further the portion of digits or numbers leftwards to the last digits, remains the same.

At that instant we use *Ekādhikena* on left digits.

Multiplication of the last digits gives the right part of the answer.

Example 1: To calculate 47 x 43

We can see that the end digits sum 7 + 3 = 10; then by the corollary *Antyayordaśake'pi* and the formula *Ekādhikena* we have the answer:

$$\begin{aligned} 47 \times 43 &= \{(4 + 1) \times 4\} / 7 \times 3 \\ &= 20 / 21 \\ &= 2021 \end{aligned}$$

Example 2: To calculate 127 x 123

As *Antyayordaśake'pi* works we apply *Ekādhikena*:

127 x 123 = 12 x 13 / 7 x 3 = 156 / 21 = 15621.

It is further interesting to note that the same rule works when the sum of the last 2, last 3, last 4 – – – digits added respectively equal to 100, 1000, 10000 —– – .

The simple point to remember is to multiply each product by 10, 100, 1000, – – as the case may be.

It can be observed that this is more convenient while working with the product of 3 digit numbers.

Example 3: To calculate 292 x 208

Here 92 + 08 = 100, Left portion is same i.e. 2
292 x 208 = (2 x 3) / 92 x 8
= 60 / 736 (for 100 raise the Left part of the product by 0)
= 60736.

Example 4: To calculate 848 x 852

Here 48 + 52 = 100, Left portion is 8 and its *Ekādhikena* is 9.
Now Right product 48 x 52 can be obtained by *Ānurūpyeṇa* mentally.

48 2
52 2
______
= 2496

and write 848 x 852 = 8 x 9 / 48 x 52
= 720 / $_2$496
= 722496. (Since left part of the product is to be multiplied by 10 and 2 to be carried over as the base is 100).

*****

26. Corollary 9 – अन्त्ययोरेव – *Antyayoreva*

– Only the Last Terms

Antya – last; *yor* – digit; and *eva*– only.

The *Upa–Sūtra* reads: अन्त्ययोरेव – *Antyayoreva*, which translated into English, simply says "Only the Last Terms". Its application and modus operandi are:

This is useful in solving simple equations of the below given type.

The type of equations are those whose numerator and denominator on the L.H.S., bearing the independent terms stand in the same ratio to each other as the entire numerator and the entire denominator of the R.H.S. stand to each other.

Let us have a look at the following example.

Example 1:

$$\frac{x^2 + 2x + 7}{x^2 + 3x + 5} = \frac{x + 2}{x + 3}$$

In the conventional method we proceed as

$$\frac{x^2 + 2x + 7}{x^2 + 3x + 5} = \frac{x + 2}{x + 3}$$

$$(x + 3)(x^2 + 2x + 7) = (x + 2)(x^2 + 3x + 5)$$

$$x^3 + 2x^2 + 7x + 3x^2 + 6x + 21 = x^3 + 3x^2 + 5x + 2x^2 + 6x + 10$$

$$x^3 + 5x^2 + 13x + 21 = x^3 + 5x^2 + 11x + 10$$

$13x + 21 = 11x + 10$ canceling like terms on both sides

$$2x = -11$$

$$x = -11 \div 2$$

Now we solve the problem using *Antyayoreva*:

$$\frac{x^2 + 2x + 7}{x^2 + 3x + 5} = \frac{x + 2}{x + 3}$$

We observe that

$$\frac{x^2 + 2x}{x^2 + 3x} = \frac{x(x + 2)}{x(x + 3)} = \frac{x + 2}{x + 3}$$

This satisfies the condition in the corollary. Hence from the corollary:

$$\frac{x + 2}{x + 3} = \frac{7}{5}$$

$$5x + 10 = 7x + 21$$
$$2x = -11$$
$$x = -11 \div 2$$

We will use the application of the corollary in another type of problem.

Example 2: $(x + 1)(x + 2)(x + 9) = (x + 3)(x + 4)(x + 5)$

Re–arranging the equation, we have

$$\frac{(x + 1)(x + 2)}{(x + 4)(x + 5)} = \frac{x + 3}{x + 9}$$

i.e.

$$\frac{(x^2 + 3x + 2x + 3)}{(x^2 + 9x + 20x + 9)}$$

$$\frac{(x^2 + 3x)}{(x^2 + 9x)} = \frac{x(x + 3)}{x(x + 9)} = \frac{(x + 3)}{(x + 9)}$$ gives the solution by *Antyayoreva*

$$\frac{(x + 3)}{(x + 9)} = \frac{2}{20}$$

$$20x + 60 = 2x + 18$$
$$20x - 2x = 18 - 60$$
$$18x = -42$$
$$\therefore \quad x = -42 \div 18$$
$$= -7 \div 3.$$

Once again look into the problem

$(x + 1)(x + 2)(x + 9) = (x + 3)(x + 4)(x + 5)$

Sum of the binomials on each side

$x + 1 + x + 2 + x + 9 = 3x + 12$
$x + 3 + x + 4 + x + 5 = 3x + 12$

It is the same.

In such a case, the equation can be adjusted into the form suitable for application of *Antyayoreva*.

Example 3: $(x + 2)(x + 3)(x + 11) = (x + 4)(x + 5)(x + 7)$

Sum of the binomials on L.H.S. –3x + 16 = Sum of the binomials on R.H.S
Hence *Antyayoreva* can be applied. Adjusting we get

$$\frac{(x + 2)(x + 3)}{(x + 4)(x + 7)} = \frac{x + 5}{x + 11} = \frac{2 \times 3}{4 \times 7} = \frac{6}{28}$$

$$28x + 140 = 6x + 66$$
$$28x - 6x = 66 - 140$$
$$22x = -74$$

$$x = \frac{-74}{22} = \frac{-37}{11}$$

*****

27. Corollary 10 – समुच्चयगुणितः – *Samuccayaguṇitaḥ*

– The Sum of the Products

Samuccaya – the whole; and *guṇitaḥ* – The properties or qualities.

The *Upa–Sūtra* reads: समुच्चयगुणितः – *Samuccayaguṇitaḥ*, which translated into English, simply says "The Sum of the Products". Its application and modus operandi are:

In connection with factorisation of quadratic expressions this *Upa–Sūtra*, viz. *Guṇita samuccayaḥ* – *Samuccaya Guṇitaḥ* is useful

.

It is intended for the purpose of verifying the correctness of obtained answers in multiplications, divisions and factorisations.

It can be interpreted more aptly in this context as: "The product of the sum of the coefficients (sc) in the factors is equal to the sum of the coefficients (sc) in the product"

Symbolically we represent as:

sc of the product = product of the sc (in the factors)

Example 1: $(x + 3)(x + 2) = x^2 + 5x + 6$

Now $(1 + 3)(1 + 2) = 1 + 5 + 6$

$4 \times 3 = 12$: Thus verified.

Example 2: $(x + 5)(x + 7)(x - 2) = x^3 + 10x^2 + 11x - 70$

$(1 + 5)(1 + 7)(1 - 2) = 1 + 10 + 11 - 70$

i.e., $6 \times 8 \times (-1) = 22 - 70$

i.e., $-48 = -48$ Verified.

The Pythagoras theorem can be proved using this corollary.

*****

28. Corollary 11 – लोपनस्थापनाभ्याम् – *Lopanasthāpanābhyām*

– By Alternative Elimination and Retention

Lopana – violate, destroy, injure, interrupt, omit; *Sthāpana* – maintain, preserve, support, establish, make stable; and *Bhyām* – use both.

The *Upa–Sūtra* reads: लोपनस्थापनाभ्याम् – *Lopanasthāpanābhyām*, which translated into English, simply says "By Alternative Elimination and Retention". Its application and modus operandi are:

This corollary and the main formula 11 – व्यष्टिसमष्टिः – *Vyaṣṭisamaṣṭiḥ* – "Specific and General" are inseparable twins and hence explained here in a combined manner.

Consider the case of factorisation of quadratic equation of type

$$ax^2 + by^2 + cz^2 + dxy + eyz + fzx$$

This is a homogeneous equation of second degree in three variables x, y, z. The *Upa–Sūtra* removes the difficulty and makes the factorisation simple. The steps are as follows:

I. Eliminate z by putting z = 0 and retain x and y and factorise thus obtained quadratic equation in x and y by means of *Ādyamādyena Sūtra* ;
II. Similarly eliminate y and retain x and z and factorise the quadratic in x and z.
III. With these two sets of factors, fill in the gaps caused by the elimination process of z and y respectively. This gives actual factors of the expression.

Example 1: $3x^2 + 7xy + 2y^2 + 11xz + 7yz + 6z^2$.

Step (i): Eliminate z and retain x, y; factorise

$$3x^2 + 7xy + 2y^2 = (3x + y)(x + 2y)$$

Step (ii): Eliminate y and retain x, z; factorise

$$3x^2 + 11xz + 6z^2 = (3x + 2z)(x + 3z)$$

Step (iii): Fill the gaps, the given expression

$$= (3x + y + 2z)(x + 2y + 3z)$$

Example 2: $3x^2 + 6y^2 + 2z^2 + 11xy + 7yz + 6xz + 19x + 22y + 13z + 20$

Step (i): Eliminate y and z, retain x and independent term
i.e., $y = 0$, $z = 0$ in the expression (E).
Then $E = 3x^2 + 19x + 20 = (x + 5)(3x + 4)$

Step (ii): Eliminate z and x, retain y and independent term
i.e., $z = 0$, $x = 0$ in the expression.
Then $E = 6y^2 + 22y + 20 = (2y + 4)(3y + 5)$

Step (iii): Eliminate x and y, retain z and independent term
i.e., $x = 0$, $y = 0$ in the expression.
Then $E = 2z^2 + 13z + 20 = (z + 4)(2z + 5)$

Step (iv): The expression has the factors (think of independent terms)
$= (3x + 2y + z + 4)(x + 3y + 2z + 5)$.

In this way either homogeneous equations of second degree or general equations of second degree in three variables can be very easily solved by applying *Ādyamādyenāntyamantyena* and *Lopanasthāpanābhyām* *Sūtra–s*.

Highest common factor:

To find the Highest Common Factor i.e. H.C.F. of Algebraic expressions, the factorisation method and process of continuous division are in practice in the conventional system. We now apply *Lopanasthāpanābhyām* *Sūtra*, the *Saṅkalana Vyavakalanābhyām* process and the *Ādyamādyenāntyamantyena* rule to find out the H.C.F in a more easy and elegant way.

Example 3: Find the H.C.F. of $x^2 + 5x + 4$ and $x^2 + 7x + 6$.

1. Factorisation method:

$x^2 + 5x + 4 = (x + 4)(x + 1)$
$x^2 + 7x + 6 = (x + 6)(x + 1)$
H.C.F. is $(x + 1)$.

2. Continuous division process.

$x^2 + 5x + 4)\ x^2 + 7x + 6\ (1$

$x^2 + 5x + 4$

$2x + 2)\ x^2 + 5x + 4\ (\frac{1}{2}x$

$x^2 + x$

$4x + 4)\ 2x + 2\ (\frac{1}{2}$

$2x + 2$

0

Thus $4x + 4$ i.e., $(x + 1)$ is H.C.F.

Lopanasthāpanābhyām process i.e. elimination and retention or alternate destruction of the highest and the lowest powers is as below:

$$\text{Subtract}\ \left\{\begin{array}{l} x^2 + 5x + 4 \\ x^2 + 7x + 6 \\ \hline -2x - 2 \\ \hline -2(x + 1) \end{array}\right.$$

i.e. $(x + 1)$ is H.C.F.

Example 4: $x^3 - 7x - 6$ and $x^3 + 8x^2 + 17x + 10$.

Now by ***Lopanasthāpanābhyām*** and *Saṅkalana Vyavakalanābhyām*

$$\text{Subtract}\ \left\{\begin{array}{l} x3 \quad\quad - 7x - 6 \\ x3 + 8x^2 + 17x + 10 \\ \hline -8x^2 - 24x - 16 \\ \hline -8(x^2 + 3x + 2) \end{array}\right.$$

Example 5: $2x^3 + x^2 - 9$ and $x^4 + 2x^2 + 9$.

By *Vedic Sūtra–s*:

Add: $(2x^3 + x^2 - 9) + (x^4 + 2x^2 + 9)$
$= x^4 + 2x^3 + 3x^2 \div x^2$ gives $x^2 + 2x + 3$ ———

(i) Subtract after multiplying the first by x and the second by 2.

Thus $(2x^4 + x^3 - 9x) - (2x^4 + 4x^2 + 18)$
$= x^3 - 4x^2 - 9x - 18$ ———

(ii) Multiply (i) by x and subtract from (ii)

$x^3 - 4x^2 - 9x - 18 - (x^3 + 2x^2 + 3x)$
$= - 6x^2 - 12x - 18$

$\div - 6$ gives $x^2 + 2x + 3$.

Thus $(x^2 + 2x + 3)$ is the H.C.F. of the given expressions.

Digital Roots:

When all the digits of a number are added up till there is only one digit is left, that digit is called the digital root of the number.

For instance the digital root of the number 5674 is 5 + 6 + 7 + 4 = 2 + 2 = 4.

It is also the same as the remainder when the number is divided by 9.

By using the process of elimination and retention we strike out the digit or digits whose sum is 9 and the remainder is the digital root –

For instance in the case of ~~5~~674

We strike out 5 and 4 since there sum is 9.

The sum of the remaining digits 6 and 7 is 13 and further the sum is 4.

*****

29. Corollary 12 – विलोकनम् – *Vilokanam*

– Mere Observation

Vilokanam – the act of looking or seeing, looking at, regarding, observing, contemplating looking for, finding out, perceiving, noticing, becoming aware of, paying attention to, studying.

The *Upa–Sūtra* reads: विलोकनम् – *Vilokanam*, which translated into English, simply says "Mere Observation". Its application and modus operandi are:

Generally we come across problems which can be solved by mere observation. But we follow the same conventional procedure and obtain the solution. But the hint behind the corollary enables us to observe the problem completely and find the pattern and finally solve the problem by just observation.

Pure observation, without judgment, is probably the most critical first step in any problem–solving effort.

Let us take the equation $x + \frac{1}{x} = \frac{5}{2}$

Without noticing the logic in the problem, the conventional process tends us to solve it in the following way:

$$x + \frac{1}{x} = \frac{5}{2}$$

$$2x^2 + 2 = 5x$$

$$2x^2 - 5x + 2 = 0$$

$$2x^2 - 4x - x + 2 = 0$$

$$2x(x - 2) - (x - 2) = 0$$

$$(x - 2)(2x - 1) = 0$$

$$x - 2 = 0 \text{ gives } x = 2$$

$$2x - 1 = 0 \text{ gives } x = \tfrac{1}{2}$$

But by *Vilokanam* i.e. observation

$x + \frac{1}{x} = \frac{5}{2}$ can be viewed as

$$x + \frac{1}{x} = 2 + \frac{1}{2} \text{ giving } x = 2 \text{ or } ½.$$

Example 1:

$$\frac{x}{x+2} + \frac{x+2}{x} = \frac{34}{15}$$

In the conventional process, we have to take L.C.M, cross–multiplication, simplification and factorisation. But *Vilokanam* gives:

$$\frac{34}{15} = \frac{9+25}{5 \text{ x } 3} = \frac{3}{5} + \frac{5}{3}$$

$$\frac{x}{x+2} + \frac{x+2}{x} = \frac{3}{5} + \frac{5}{3}$$

gives

$$\frac{x}{x+2} = \frac{3}{5} \text{ or } \frac{5}{3}$$

$5x = 3x + 6$ or $3x = 5x + 10$
$2x = 6$ or $-2x = 10$
$x = 3$ or $x = -5$

Simultaneous Quadratic Equations:

Example 2: $x + y = 9$ and $xy = 14$.

We follow in the conventional way that

$(x - y)^2 = (x + y)^2 - 4xy = 9^2 - 4\ (14) = 81 - 56 = 25$
$x - y = \sqrt{25} = \pm 5$
$x + y = 9$ and $x - y = 5$ gives $x = 7, y = 2$ or $x = 2, y = 7$

But by *Vilokanam*, $xy = 14$ gives $x = 2, y = 7$ or $x = 7, y = 2$

These two sets satisfy x + y = 9 since 2 + 7 = 9 or 7 + 2 = 9. Hence the solution.

Partial Fractions:

Example 3: Resolve

$$\frac{2x + 7}{(x + 3)(x + 4)} \text{ into partial fractions.}$$

We write

$$\frac{2x + 7}{(x + 3)(x + 4)} = \frac{A}{(x + 3)} + \frac{B}{(x + 4)}$$

$$= \frac{A(x + 4) + B(x + 3)}{(x + 3)(x + 4)}$$

$$2x + 7 = A(x + 4) + B(x + 3).$$

Put $x = -3$, $2(-3) + 7 = A(-3 + 4) + B(-3 + 3)$

$$1 = A(1) \quad \therefore A = 1.$$

$x = -4$, $2(-4) + 7 = A(-4 + 4) + B(-4 + 3)$

$$-1 = B(-1) \quad \therefore B = 1.$$

Thus $\frac{2x + 7}{(x + 3)(x + 4)} = \frac{1}{(x + 3)} + \frac{1}{(x + 4)}$

But by *Vilokanam* $\frac{2x + 7}{(x + 3)(x + 4)}$ can be resolved as

$(x + 3) + (x + 4) = 2x + 7$, directly we write the answer.

Example 4:

$$\frac{3x + 13}{(x + 1)(x + 2)}$$

from (x + 1) (x + 2) we can observe that

$$10(x + 2) - 7(x + 1) = 10x + 20 - 7x - 7 = 3x + 13$$

$$\text{Thus } \frac{3x + 13}{(x + 1)(x + 2)} = \frac{10}{x + 1} - \frac{7}{x + 2}$$

*****

30. Corollary 13 – गुणितसमुच्चयः समुच्चयगुणितः – *Guṇitasamuccayaḥ Samuccayaguṇitaḥ*

– The product of the sum of the coefficients in the factors is equal to the sum of the coefficients in the product

Guṇita – properties or qualities; *samuccayaḥ* – the sum (totality, final result) or quality of the total or total of the qualities.

The *Upa–Sūtra* reads: गुणितसमुच्चयः समुच्चयगुणितः – *Guṇitasamuccayaḥ Samuccayaguṇitaḥ*, which translated into English, simply says "The product of the sum of the coefficients in the factors is equal to the sum of the coefficients in the product". Its application and modus operandi are:

In connection with factorisation of quadratic expressions the *Upa–Sūtra*, viz. *Guṇitasamuccayaḥ Samuccayaguṇitaḥ* is useful. It is intended for the purpose of verifying the correctness of obtained answers in multiplications, divisions and factorisations. It means in this context:

The product of the sum of the coefficients Sc in the factors is equal to the sum of the coefficients Sc in the product.

Symbolically we represent as Sc of the product = product of the Sc (in the factors)

Example 1: (x + 3) (x + 2) = x2 + 5x + 6

Now (x + 3) (x + 2) = 4 x 3 = 12: Thus verified.

Example 2: $(x - 4)(2x + 5) = 2x^2 - 3x - 20$

Sc of the product 2 – 3 – 20 = – 21

Product of the Sc is (1 – 4) (2 + 5) = (–3) (7) = – 21. Hence verified.

In case of cubics, bi–quadratics also the same rule apply.

We have $(x + 2)(x + 3)(x + 4) = x^3 + 9x^2 + 26x + 24$

Sc of the product = 1 + 9 + 26 + 24 = 60

Product of the Sc is (1 + 2) (1 + 3) (1 + 4) = 3 x 4 x 5 = 60. Hence Verified.

Example 3: $(x + 5)(x + 7)(x - 2) = x^3 + 10x^2 + 11x - 70$

$(1 + 5)(1 + 7)(1 - 2) = 1 + 10 + 11 - 70$

i.e., $6 \times 8 \times -1 = 22 - 70$

i.e., $-48 = -48$ Verified.

We apply and interpret So and Sc as sum of the coefficients of the odd powers and sum of the coefficients of the even powers and derive that So = Sc gives (x + 1) is a factor for thee concerned expression in the variable x. Sc = 0 gives (x – 1) is a factor.

*****

31. Corollary 14 – ध्वजाङ – *Dvajāḍa* or *Dvajāṅka*

– On the flag

Dvaja = flag. The *Upa–Sūtra* reads: ध्वजाङ – *Dvajāḍa*, which translated into English, simply says "On the flag". Its application and modus operandi are:

Points of change – by observation, we can identify key points where change occurs. It is important to create mechanisms to 'flag' these points or provide indications or signals of change.

Points of importance – During the course of problem solving, while allowing the mind to diverge and work in a broad area, it is important to 'flag' or mark key points along the way.

For instance, parking a promising idea and returning to explore further is a way of marking a key breakthrough. Effective 'flagging' can ensure that nothing of promise is inadvertently forgotten in the attempt to explore wider avenues of problems and solutions.

Example 1: To calculate 716769 ÷ 54.

The trick is to reduce the divisor to a mentally manageable value by putting its other digits "on top of the flag". In this example, the divisor will be reduced to 5 (instead of 54) by pushing the 4 up the flag–post, as shown below. Corresponding to the number of digits flagged on top (in this case, one), the rightmost part of the number to be divided is split to mark the placeholder of the decimal point or the remainder portion.

716769 ÷ 54 = 13273.5

		2	2	4	4	3	20
4							
5	7	1	6	7	6	9	
	1	3	2	7	3	5	

1 7 ÷ 5 = 1 remainder 2. Put the quotient 1, the first digit of the solution, in the first box of the bottom row and carry over the remainder 2

2 The product of the flagged number (4) and the previous quotient (1) must be subtracted from the next number (21) before the division can proceed. 21 – 4 x 1 = 17

17 ÷ 5 = 3 remainder 2. Put down the 3 and carry over the 2

3 Again subtract the product of the flagged number (4) and the previous quotient (3),

26 – 4 x 3 = 14
14 ÷ 5 = 2 remainder 4. Put down the 2 and carry over the 4

4 47 – 4 x 2 = 39
39 ÷ 5 = 7 remainder 4. Put down the 7 and carry over the 4

5 46 – 4 x 7 = 18
18 ÷ 5 = 3 remainder 3. Put down the 3 and carry over the 3

6 39 – 4 x 3 = 27. Since the decimal point is reached here, 27 is the raw remainder. If decimal places are required, the division can proceed as before, filling the original number with zeros after the decimal point.

27 ÷ 5 = 5 remainder 2. Put down the 5 (after the decimal point) and carry over the 2

7 20 – 4 x 5 = 0. There is nothing left to divide; hence this clearly completes the division.

*****

32. Vedic Mathematics – A Management Perspective

Individually the usage, explanation and examples of the 16 *Sūtra*–s and 14 *Upa–Sūtra*–s have been discussed in the earlier chapters. Out of them some apparently do not convey any meaning while some seem to be ambiguous. Fourteen formulae are very specifically Mathematical and are therefore otherwise not much used in general. The Principles of *Vedic* Mathematics are recommended to be applied to solve Mathematical problems. A deeper look into the principles reveals that these are generic in nature and provide directions of thought effective in solving all types of problems. Sixteen generic principles are extracted, clustered and consolidated into seven broad directions of thought – Observation, Division (Segmentation), Equation (Comparison), Addition, Subtraction, Variation and Rotation. These principles can be used stand–alone or in combinations to provide a rich set of triggers or thought directions. These triggers are fit to be used at all stages of the problem resolution cycle right from understanding a problem to creation of ideas and solutions. The use of inventive principles or triggers stimulates creativity. These principles are widely accepted in modern day Management Science.

Most of the standard text books on *Vedic* Mathematics normally combine all the formulae into the above 7 headings and try to explain the Mathematical operations through examples or otherwise. However, here the above 7 are not treated as Mathematics (group of formulae), but as Management principles and deliberated below. This gives a perspective different from that of the monotonous thinking that these formulae can be used only in Mathematics.

Edward De Bono has proposed the use of random words as triggers to generate new directions of thought. The 40 inventive principles of TRIZ[2] are widely used as triggers to generate new ideas when resolving specific types of contradictions. The seven SCAMPER (Substitute, Combine, Adapt, Modify, Put to use, Eliminate, Rearrange) principles are useful in the context of brainstorming and ideation. In this context, it is interesting to look at the *Vedic* Mathematics principles through an "inventive triggers" lens. These principles can very well be compared to the modern day SCAMPER principles

---

[2] TRIZ is a romanised acronym for Теория решения изобретательских задач ("") meaning "The theory of solving inventor's problems" or "The theory of inventor's problem solving". It was developed by a Soviet engineer and researcher Genrich Altshuller and his colleagues in 1946.

of Management. At the outset, it does seem that the *Vedic* Principles mirror the simple techniques the human brain uses to get to solutions in a systematic manner. Experiments with these principles in live brainstorming and ideation sessions have proved fruitful.

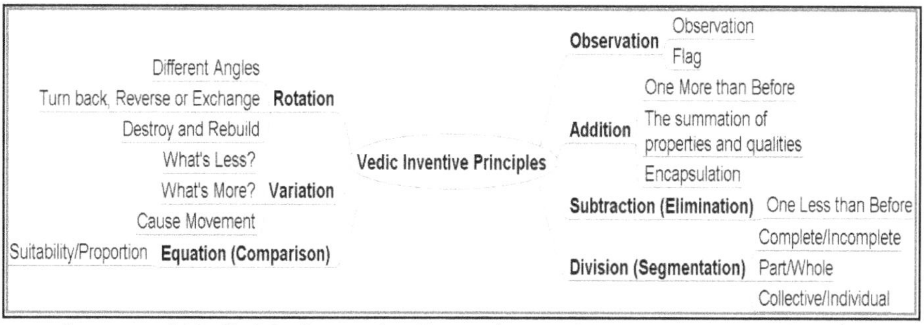

Groups of *Vedic* Mathematics Formulae and Management Principles

The use of triggers to stimulate new thoughts is a well–documented beneficial practice. Triggers can be random and infinite (use of any word), or specific and finite. These formulae also triggers or thought directions in contexts of ideation, brainstorming and problem solving.

Sixteen of these triggers (as described in the above diagram) have been consolidated into seven broad directions of thought:

1. **Observation** is critical to create a broad perspective and open up multiple avenues for exploration.
2. **Division (Segmentation)** helps slice a scenario in multiple ways.
3. **Addition** – looking at adding, merging or combining elements to create something new.
4. **Subtraction (Elimination)** – looking at a perspective of removing or eliminating parts of the system in question.
5. **Variation** – observing and causing change.
6. **Rotation** – looking at ways to re–orient the problem to create new perspectives.
7. **Equation (Comparison)** – the ability to match, compare and choose.

32.1 Observation

***Vilokanam* – Observation;** Pure observation, without judgment-unbiased, is probably the most critical first step in any problem–solving effort.

- Opening the mind to become receptive to data is of utmost importance.
- Focused observation can facilitate broader understanding and prevent narrow views.
- Observation (going and seeing oneself) is one of the key principles of Toyota's "Lean Thinking".
- "Pure observation" or "White Hat" thinking is usually the first step in a Six Thinking Hats brainstorming session as espoused by Edward De Bono.

***Dvajam* – Flag**

- Points of change – By observation, one can identify key points where change occurs. It is important to create mechanisms to 'flag' these points or provide indications or signals of change.
- Points of importance – During the course of problem solving, while allowing the mind to diverge and work in a broad area, it is important to 'flag' or mark key points along the way. For instance, parking a promising idea and returning to explore further is a way of marking a key breakthrough. Effective 'flagging' can ensure that nothing of promise is inadvertently forgotten in the attempt to explore wider avenues of problems and solutions.
- Flagging is an important concept in computer programming. For instance systems flag the records already processed or that are not to be processed so that duplicate processing is avoided.

32.2 Division (Segmentation)

Slice a scenario in multiple ways - ***Vyaṣṭisamaṣṭiḥ* – Part and whole;** Constituents of this concept can be:

- Divide an object into constituent parts
- Divide a transaction into constituent actions
- Divide a scenario into objects, people and actions
- Divide a day into hours
- Divide a context into facts and perceptions

Super–system:

- The 10,000 – 20,000 – 50,000 feet views.
- The whole is greater than the sum of parts – looking at system behaviour

which is manifested only in the whole and not in the parts
  - The molecular strength of Carbon–60 (Carbon–60 is one of the strongest molecules, but individually Carbon is very light)
  - Foraging behaviour of ants
- The whole is also a part of a greater whole – every super–set is a sub–set of another super–set.

In any Project/ Program Management, especially in computer systems, in planning process, slicing the entire project into activities, tasks, etc., is a very important area to focus on, failing which the entire project is likely to fail.

Vyaṣṭisamaṣṭiḥ – Collective and individual

Collective

- View objects as a collective unit rather than as individual units
  - The utility of cars in general as opposed to the utility of a specific model
  - Market trends
- Evaluate collective behaviour
  - Teaming strategy
  - Mob mentality

Individuality

- Focus on one object as an individual entity and evaluate its interactions with everything around it
- Focus on one perspective individually at a time – Six Thinking Hats (White: neutral, objective, Red: emotional, angry, Black: serious, somber, Yellow: sunny, positive, Green: growth, fertility and Blue: cool, sky above)
- Associate actions with a specific person rather than with a generic profile – Ram likes to eat chocolate vs. Boys like to eat chocolate

***Pūraṇāpūraṇābhyām* – Complete and Incomplete**

The human brain reduces complexity by forming patterns. Over time, some patterns become fixed or rigid. Grey gets sorted into either the black or the white box. The tendency of the brain is to 'complete' the pattern quickly. Since this happens subconsciously, it can be difficult to identify, 'when' this happens. While forming patterns, the brain compensates for both missing data as well as extra data. Any data that does not "fit in" can get subconsciously discarded. It is important therefore to take a deeper look to identify the difference between perception and reality of what is 'complete' and what is 'incomplete'.

32.3 Addition

Add, merge, combine or increase to create something new – **One More – *Ekādhikena Pūrveṇa* – One more than before;** Add an object/ Combine objects

- Swiss knife
- Tooth–brush with tongue cleaner on the back surface
- Cell phones with camera
- Vacuum cleaner with dustbin

Merge functions so that a separate object is not needed

- Board for chopping, grating, dicing vegetables
- Car battery charges while the car is running
- Pollination happens while the bee collects nectar

Summation of Properties and Qualities

***Guṇitasamuccayaḥ* – The sum of properties and/ or**
***Guṇakasamuccayaḥ* – The sum of qualities**

Identify all the properties and qualities of the system for e.g. length, strength, colour, efficiency, cost, etc. Rather than looking at one property in isolation, look at the summation of the properties say length and colour, or strength, efficiency and cost.

- As Lean Thinking suggests measure higher rather than lower. E.g. measuring the 'wear' of a tyre combines the measurement of material strength, distance traveled, road conditions, average speed and frequency of rotation.
- Improve multiple parameters at once rather than one at the cost of other or arriving at middle ground. E.g. decrease weight + increase strength + decrease cost.

Samuccayaḥ Guṇitaḥ – The property of the sum or whole

The whole is greater than the sum of parts – look at system behaviour which is manifested only in the whole and not in the parts.

- The molecular strength of Carbon–60 (Carbon–60 is one of the strongest molecules, but individually Carbon is very light)
- Foraging behaviour of ants
- Volume is created only when length, breadth and width combine

Encapsulation – *Veṣṭanam* – Surround, cover or enclose.

- Add a layer to hide the details of the system
- Add a protective layer or substance
- Create a layer of abstraction
- Convert part of the system into a black box

32.4 Subtraction

Remove, eliminate, reduce or decrease - *Ekanyūnena Pūrveṇa* – One less than before

Remove a resource

- How a building can be constructed in 'one-less' day
- How a boat can be rowed with one less person

Remove a constraint

- If cost is not a constraint, can we have a different solution?
- If the lock is not to have a key, how will the function be achieved? – number-lock

Decrease/ reduce (many–to–one or more–to–less)

- Decrease the number of objects performing the same function
  - Table with 4 legs – table with three legs – two broader legs – one cylindrical leg?
  - Number of redundant keys on the keyboard

- Remove/ reduce objects with overlapping functions
  - Ceiling fans in a well–ventilated space
  - Fans in rooms with Air–conditioners
- Reduce harmful effects
  - How to decrease the rate of deflation of a punctured tyre – leading to tubeless tyres.

Eliminate an object that is not contributing to function

- Appendix in the human body

This concept (Remove, eliminate, reduce or decrease) is an important chapter in Project/ Program Management mainly in computer systems to reduce cost and time.

32.5 Variation

Observe and create change - What is more? –

***Śeṣāṇyaṅkena Carameṇa* – The sum of what's left over**

- Identify things that are extra or in excess – why are these in excess?
- Identify things that are unutilised – how can they be used?
- Identify things that are left over or are by–products – how can this be re–used?
- Identify points of improvements in performance – what is causing the variation?

What's less? - ***Yāvadūnam* – By whatever is less**

- Identify things that are not available in adequate quantity – gaps in the system.
- Identify dips in performance – what is causing a variation?
- Identify things that are borrowed from other parts of the system – what is missing in this part of the system that has to be covered by other parts?
- Identify delays – what is causing inadequacy of time?
- Identify points of stress or duress – what is missing that causes this stress?

Cause Movement - ***Calana–kalanābhyām* – Set in motion or cause change**

- Create movement in anything stationary – objects, parameters, thoughts
- If movement is the norm, try becoming stationary
- Change anything that is constant
  - Engines rotating at constant speed – drive at different speeds
  - Processes that are unchanged over a long period of time – introduce continuous variations
  - Personal habits, say exercise, use different combinations everyday
- Random changes by choice – genetic algorithm

32.6 Rotation

Reorient to create new perspectives - ***Parāvartya Yojayet* – Turn back, exchange or reverse**

Reverse

- Rather than looking at how to make it work look at how to make it fail
- Cup – half empty or half full?
- Instead of jogging fast jog slowly
- Move the bell rather than the gong
- Road runs backward instead of you running forward – treadmill, escalator
- Water faucet – tap mouth upwards rather than downwards
- Code first – design later – iteratively

Exchange, Substitute, Replace

- Manager and subordinate exchange roles for a week to understand each other's job pressures.
- Enter digits first, dial and connect later
- Replace expensive items with inexpensive objects achieving the same function – refilling toner of a printer instead of replacing it.

Different Angles

***Ūrdhva–Tiryagbhyām* – Vertically and horizontally**

Change the perspective

- Depth–first rather than breadth–first and vice–versa
- Bottom–up rather than top–down and vice–versa
- Town–planning – rather than viewing it at ground–level, how would an aerial view look like?
- How about a different cultural perspective?
- Approach a problem from the end rather than the start (or from the middle?)
- Look at things you don't usually look at – how does a car look from below?
- Look at things from the side – neutrally or passively

Consider a new dimension

- Linear – planar – 3D – Space (4D) – Time (5D)
- Lines – curves
- Degree of freedom – robotic arm, Japanese martial art segmented stick
- Analog – Digital
- Sound – Light – Heat

***Lopanasthāpanābhyām* – Destroy and Rebuild**

Often, to break out of a dead–end of repetitive patterns, it is important to destroy the existing patterns, clear the mind and rebuild from scratch. The same approach can be used while designing systems where first–cut designs can be dismantled and rebuilt from scratch. Sometimes, rather than continue to improve existing systems through patchwork solutions, it might be better to rebuild from scratch.

Destroy, disrupt and interrupt

- Systematically destroying a system can be a good way to detect faults (and strengths) in the system (subversion analysis).
- Interrupting a system can help identify points of inertia.
- Ideas to break existing systems often lead to the most innovative ideas to improve or create new systems.
- Routine random disruptions help systems evolve mechanisms to recover and thereby become more robust.

Destroy and Rebuild

- Re–factoring of systems involves the systematic destruction and rebuilding of systems on a part–by–part basis.
- This phenomenon is also seen regularly as part of natural processes – the cycles of death and birth of systems including living organisms e.g. evaporation – rainfall, forest fires – fertile soil etc.

32.7 Equation (Comparison)

Match, compare and choose - **Suitability/ proportion**

***Ānurūpye Śūnyamanyat* – Everything else, other than what is in proportion or is suitable, is zero or absent.**

***Ādyamādyena Antyamantyena* – first by first and last by last**

Compare apples with apples and oranges with oranges.

- Nail and hammer, screw and screwdriver.
- Cotton in summer, wool for winter.
- For efficiency of operation, generic processes are tailored, so that they become suitable for use in specific contexts. For instance – customisation of a global software product to the clients' specific needs.

***Ānurūpyeṇa* – In Proportion**

- Increase in temperature – increase in sale of ice–creams
- Increase in number of snakes – reduction in number of rodents – increase in volume of crop harvested

Comparison/ Equation

- Comparing with something similar
- Comparing with something dissimilar
- Drawing parallels/ equate

Inertia of familiarity

Interestingly, the principle also points out that the human brain actively looks for suitability or proportion – familiar patterns. When encountering a problem, one can be hemmed in by a pet solution which blanks out all other possibilities. In this way, this principle is also a warning to actively avoid the familiarity trap. (This perspective can be generated by applying the 'Reverse' principle on this principle itself)!

32.8 Summary

The triggers stimulate the thought of "differently thinking", "diagonally thinking", "vertically thinking", etc. All these thought processes are needed for effective Management. All these are used widely and with a purpose in computer Project/ Program Management also.

*****

33. Criticism on Vedic Mathematics

Like any upright entity in the world, *Vedic* Mathematics also is not beyond the sight of critics. In Tamil there is a proverb "only the tree with ripe fruits will be hit by stones". However, one thing to be observed is, no critic has questioned the working of any of the formulae. But have instead questioned the very name "*Vedic* Mathematics". This is the standard criticism attempted by most.

The *Sūtra–s* are claimed to be a set of Mathematical principles rediscovered from ancient Hindu texts of knowledge (the *Vedas*). This has become a controversial assertion – many believe that these principles are neither *Vedic* nor related to Mathematics.

Critics have questioned whether this subject deserves the name *Vedic* or indeed Mathematics with any other prefix or suffix. They point to the lack of evidence of any *Sūtra–s* from the *Vedic* period consistent with the system, the inconsistency between the topics addressed by the system (such as decimal fractions) and the known Mathematics of early India, the substantial extrapolations from a few words of a *Sūtra* to complex Arithmetic and the restriction of applications to convenient cases. They further say that such Arithmetic as is speeded up by application of the *Sūtra–s* can be performed on a computer or calculator anyway, making their knowledge rather irrelevant in the current world.

They are also worried that it deflects the attention from genuine achievements of ancient and modern Indian Mathematics and Mathematicians and that its promotion by Hindu nationalists may damage Mathematics education in India.

There has been much controversy among Indian scholars about *Swāmijī*'s claims that the Mathematics is *Vedic* and that it encompasses all aspects of Mathematics. First, *Swāmijī*'s description of the Mathematics as *Vedic* is most commonly criticised on the basis that, thus far, none of the *sūtra*–s can be found in any extant *Vedic* literature. However, trying to locate *Swāmijī*'s references in the *Vedic* literature would be extremely difficult as it is possible that *Swāmijī* rediscovered and reconstructed the *sūtra*–s from stray references scattered throughout the *Atharva Veda*, making it difficult to trace them.

In response to criticisms that the *sūtra*–s cannot be located within the texts, several people have explained how textual references should not be the basis for evaluating the *Vedicity* of the Mathematics. Some propose that *Vedic* Mathematics is different from other Scientific work because it is not pragmatically worked out, but is based on a direct revelation or an "intuitional visualisation" of fundamental Mathematical truths. *Swāmijī* has been described as having the same "reverential approach" towards the *Vedas* as the ancient sages that formed them. Thus, it seems as though some believe that *Swāmijī* may not have found the *sūtra*–s within the *Vedas*, but that he received them spiritually as the sages did, which should validate them as *Vedic*.

The controversy about the *Vedicity* of the Mathematics is further confused by the double meaning of *Veda*. Since *Veda* can be translated to mean 'knowledge', it is also possible that *Vedic* Mathematics simply refers to the fact that the *sūtra*–s are supposed to present all knowledge of Mathematics. *Swāmijī*'s definition of *Veda* does not clearly clarify whether he uses it to represent "all knowledge" or the *Vedic* texts; rather, it seems that he uses it to refer to both.

Considering the lack of references to the *sūtra*–s, coupled with the fact that the language style does not seem *Vedic*, some propose that the *sūtra*–s were simply composed by *Swāmijī* himself. In that case, one must consider what motivated *Swāmijī* to attribute the Mathematical *sūtra*–s to the ancient texts. Was it because they are from the *Vedas*, or does claiming so give them more credibility?

Other areas of controversy regarding *Vedic* Mathematics focus on the actual Mathematics itself. *Swāmijī*'s assertion that the 16 *sūtra*–s of *Vedic* Mathematics encompass all branches of Mathematics is an extreme one, even if true and so it is not surprising that many Mathematicians challenge it. They point to the inconsistency between the topics addressed by the system (such as decimal fractions) and the known Mathematics of early India, the substantial extrapolations from a few words of a *sūtra* to complex Arithmetic strategies and the restriction of applications to convenient, special cases. They further say that such arithmetic as is sped up by application of the *sūtra*–s can be performed on a computer or calculator anyway, making their knowledge rather irrelevant in the modern world.

They are also worried that it deflects attention from genuine achievements of ancient and modern Indian Mathematics and Mathematicians.

33.1. Some replies to the criticisms

Though there is much positivity and much to look forward to, there is one important area that needs to be clarified and that is the answer to the question: "What is meant by '*Vedic* Mathematics': what is included within this term and what is not"?

Different people may have different ideas about the answer to this question. And it is not enough to simply refer to the book by *Swāmijī* as new material is being produced and termed *Vedic* Mathematics all the time.

How do we decide if a particular piece of Mathematics is *Vedic* Mathematics or not? What is the defining characteristic(s) of *Vedic* Mathematics by which we can recognise something as part of the *Vedic* system?

Why is this question important? It is vital that this question is answered because there need to be some clear boundary to what *Vedic* Mathematics is and what is not? Otherwise people can declare all sorts of obscure Mathematical results and claim them to be *Vedic* Mathematics.

Another reason why the question is important is that for many people *Vedic* Mathematics has become synonymous with tricks, short cuts and fast methods. This is unfortunate as it means it is not seen seriously considered by Mathematicians and educationists and it entirely misses the comprehensive and complete nature of the system of *Vedic* Mathematics.
Here are some possible answers to the question: what characteristic of a method, proof etc., makes it *Vedic*?

It can be construed that a method, proof etc., is *Vedic* if:

1 It comes under one or more of the *Vedic Sūtra*–s
2 It follows a method given by *Swāmijī*
3 It is a one–liner

Answer 1 – The criticisms may be rejected since the *Sūtra*–s describe natural ways in which the mind works, hence any way of thinking or any method must use the *Sūtra*–s. Even the current long multiplication method (that cannot be described as *Vedic*) uses these *Sūtra*–s.

Answer 2 – Does not allow the possibility of methods being *Vedic* that are not given by *Swāmijī* and hence is too restrictive.

Answer 3 – The word *Veda* comes from the root *Vid*, which means to know. Hence *Veda* means knowledge. These are direct intuitional revelations of the eternal truths that got revealed by itself to the great ancient sages, without any author in particular. They are without beginning and end – hence, the *Vedas* are *apourșeya* and *anādi* – not created by humans and eternal. All these attributes hold good for *Vedic* Mathematics also. Hence it is apt in calling this as *Vedic* Mathematics. *Swāmijī* himself claims that he did not write this, but he compiled the formulae as *Vedas* were compiled by sage *Vyāsa*.

Answer 4 – The reply is clear from the title of the book itself by the *Swāmijī* –

"*Vedic* Mathematics
Or Sixteen Simple Formulae from the *Vedas*
(For one–line Answers to all Mathematical problems)"

This is the full title: first *Vedic* Mathematics, then an alternative title and then the words in brackets.

This title implies that the *Sūtra*–s are the basis for *Vedic* Mathematics and that they give one–line answers to all Mathematical problems.

This is perhaps the criterion: a method must be one–line. But this answer is no use to us unless we can say what 'one–line' means.

In the area of computation we can say 'one–line' means that the answer can be given digit by digit with an occasional carry digit(s), which can be held in the mind. The term also suggests that the flow of attention is one–line: that the attention is not fragmented.

A good illustration is perhaps the obtaining of the product of two numbers by the "Vertically and Crosswise" method and by the usual conventional method. In the *Vedic* method the attention moves through the numbers being combined, obtaining the digits of the answer one after the other (from right to left or from left to right) using a simple pattern.

The conventional method is fragmented, obtaining first one row of figures and then another until finally these rows are added up. In fact the same number of products is found in both methods, but the *Vedic* method flows in one direction: i.e. it is one–line.

Getting an answer digit by digit with an occasional carry digit also suggests that such a calculation can be carried out mentally and in fact *Swāmiji* writes in his explanatory exposition: "...by means of what we have been describing as straight, single–line, mental arithmetic", suggesting that 'one–line' is equivalent to 'mental'.

Hence it can therefore be proposed that the main criterion for a method to be *Vedic* is that it has this one–line feature. But a technique should not be rejected outright on the basis that it is not one–line. If it is better than the conventional method, it is worthy of consideration with a view to further development.

The illustration of multiplication happens to be a convenient one. Mathematics is not just about computation though, but about proof, problem solving, structure and so on. So how does the one–line flow of attention criterion apply more generally? This has to be debated and decided.

Of all the four *Vedas* put together, out of 1131 *śākās*, we presently have only 11 branches. Who knows – these *Vedic* Mathematics formulae might form part of the missed branches of *Vedas*. This could have come to the mind of *Swāmijī* on account of the powers of his penance.

Without a clear idea of the boundaries of *Vedic* Mathematics, the subject may gradually disappear, with only certain techniques which are popular, remaining but not being known to be part of a complete system of *Vedic* Mathematics.

One shortcoming, if not criticism, about *Vedic* Mathematics is – the formulae are cryptic and have to be appropriately interpreted. Unlike current day Mathematics, for a similar type of problem we may have to use different formulae depending on various cases. For instance to find a square of a number, if the number ends with 5, one formula has to be applied; else another formula and so on. Hence, unless a thorough knowledge of all the formulae is had, it is difficult to apply. This is one of the main reasons that *Vedic* Mathematics is not very popular.

Let all these criticisms be there; the glory of *Vedic* Mathematics cannot be un–credited. Ignoring all these, people will continue their study. In this process, each of the formulae and corollaries had been explained in brief in the above chapters.

The last 10 years have seen a huge increase in interest in *Vedic* Mathematics. The system reconstructed by *Swāmijī* almost a century ago, known as *Vedic* Mathematics – made known to the public around 50 years back – is at last being recognised as having tremendous potential in all sorts of areas: educational, computational, scientific, psychological and so on. This century is sure to see this continue, develop and expand further.

The triggers stimulate the thought of "differently thinking", "diagonally thinking", "vertically thinking", etc. All these thought processes are needed for effective Management. All these are used widely and with a purpose in computer Project/ Program Management also.

33.2. Summary

The sixteen formulae correspond to sixteen vowels of Saṁskṛit language. With vowels all the words are formed and without which no word can be designed. Similarly any problem of Mathematics can be solved using one or more of the *Vedic* Mathematics formulae.

Vedic Mathematics is itself called as "Mental Mathematics". This gives lot of work to the brain and the mind – leading to various thinking processes.

Vedic formulae not only tell us how to do all the Mathematical calculations by easy one line method and through rapid processes, but they also tabulated the results in the shape of special corollaries containing merely illustrative specimens with a master–key for "unlocking other portals" too.

*****

Annexure 1

Brief biography of ***HH Jagadguru Swāmi Śrī Bhārati Kṛṣṇa Tīrthaji Mahāraj*** (जगद्गुरु स्वामि श्री भारती कृष्ण तीर्थजी महाराजः March, 1884 – February 2, 1960)

Early life

Venkatrama Shāstri, the *pūrvaśrama* name of the *swāmiji*, was born in March, 1884 to P. Narasimha Shāstri, originally the Tahsildar at Tirunelveli in Madras Presidency. Narasimha Shāstri later became the Deputy Collector of the Presidency. Venkatraman was born in a highly illustrious family. His uncle, Chandrasekhara Shāstri was the Principal of the Maharaja's College in Vizianagaram, while his great–grandfather, Justice C. Ranganath Shāstri was a judge in the Madras High Court.

Educational career

Venkatrama Shāstri started his educational career as a student of the National College at Tiruchi. After that he moved to the Church Missionary Society College and eventually the Hindu College, both in Tirunelveli. He was consistently securing the first place in all subjects in all of his classes. Shāstri passed his matriculation examination from the Madras University in January, 1899, where also he finished as the top of the class.

As a student Venkatraman was marked for his splendid brilliance, superb retentive memory and an insatiable curiosity. By deluging his teachers with piercing questions, making them uneasy and frequently forcing them to admit ignorance he was considered a terribly mischievous student.

Although Venkatraman always scored high in subjects like Mathematics, Sciences and Humanities, he was also proficient in languages and particularly adept in Saṁskṛit. According to his own testimonials, Saṁskṛit and oratory were his favorite subjects. Such was his mastery over the language, that he was awarded the title *'Saraswati'* by the Madras Saṁskṛit Association in July, 1899 at the age of 16. At about that time, Venkatraman was profoundly influenced by his Saṁskṛit *guru Śrī Vedam* Venkatrai Shāstri, whom he remembered with deepest love, reverence and gratitude, with tears in his eyes.

Venkatraman won the highest place in the graduation – B.A. examination in 1902. He then appeared for the M.A. Examination from the American College of Sciences, in Rochester, New York from the Bombay center in 1903. He passed the M.A. examination in seven subjects that he had chosen – Saṁskṛit, Philosophy, English, Mathematics, History, Science and another subject – simultaneously scoring the highest honours in all, which is perhaps an all–time world record, yet to be broken.

Venkatrama Saraswati, as he was called after receiving the title, also contributed to W. T. Stead's "Review of Reviews" on topics as diverse as religion and Science. During his college days, he also wrote extensively on History, Sociology, Philosophy, Politics and Literature. Reading of the latest scientific research and discoveries was his hobby throughout his life.

Early Public Life

Venkatrama Saraswati worked under Gopala Kṛśna Gokhale in 1905 for the National Education Movement and the South African Indian issues. However, his inclination towards Science and Indic studies led him to study the ancient Indian Holy Scriptures, *Adhyātma–Vidya*. In 1908 he joined the Śringeri Mutt in Mysore to study under the Śringeri *Śaṅkarācārya Śrī Satchitānanda Śivābhinava Nṛsimha Bhārati Swami*. However, his spiritual orientation was interrupted when he was called by nationalist leaders to head the newly started National College at Rajmahendri. Prof. Venkatrama Saraswati taught at the college for three years. But in 1911, he suddenly left the college to go back to *Śrī Satchitānanda Śivābhinava Nṛsimha Bhārati Swami* at the Śringeri Mutt in his quest for spiritual knowledge.

Spiritual Path

Returning to Śringeri, Venkatraman spent couple of years studying advanced *Vedānta* philosophy at the feet of *Śrī Satchitānanda Śivābhinava Nṛsimha Bhārati Swami*.

He also practiced vigorous meditation, *Brahma–sādhana* and *Yoga–sādhana* during those years in the nearby forests. It is believed that he attained spiritual self–realisation during these years in the Śringeri Mutt. He would leave the material world and practice *yoga* meditation in seclusion for many days. During those years, he also taught Saṁskṛit and Philosophy to local schools and *āshrams*. He delivered a series of lectures on *Śaṅkarācārya*'s philosophy at *Śaṅkara* Institute of Philosophy, Amalner (Khandesh). During that time, he also lectured as a guest professor at various institutions in Bombay, Puna and Khandesh.

Initiation into *Sanyāsa* order

After Venkatraman's spiritual practice and study of *Vedānta* and *Vedic* philosophy, he was initiated into the holy order of *Sanyāsa* at Benares by *Jagadguru Śaṅkarācārya Śrī Trivikram Tīrthaji Maharāj* of *Shāradāpeeth* on July 4, 1919 and on this occasion he was given the title of *Swāmi* and the new name, *"Swami Bhārāti Kṛṣṇa Tīrtha"*.

Śaṅkarācārya of *Sharāda Peeta*

Swāmiji was installed as *Śaṅkarācārya* of *Śārada Peetha* in 1921 after just two years of *Sanyāsa*. After assuming the pontificate, he was given another title, *Jagadguru*, as is the tradition. The *Swāmiji* then toured India from corner to corner giving lectures on *Sanātana Dharma*, *Vedic* philosophy and *Vedānta*. By his scintillating intellectual brilliance, powerful oratory, magnetic personality, sincerity of purpose, indomitable will, purity of thought and loftiness of character he took the entire intellectual and religious class by storm.

Śaṅkarācārya of *Govardhan Pīṭha*

Around the time the *Swāmiji* became the *Śaṅkarācārya* of *Shārada Peetha*, the *Śaṅkarācārya* of *Govardhan Pīṭha,* Puri, *Jagadguru Śaṅkarācārya Śrī Madhusudhan Tirtha*, was in failing health and was greatly impressed by our *Swāmiji*. *Madhusudan Tirtha Swāmi* requested *Swāmiji* to succeed him at the *Govardhan Pīṭha*; however *Swāmiji* respectfully declined the offer. Again, in 1925, *Śaṅkarācārya Śrī Madhusudhan Tirtha*'s health took a serious turn and *Swāmiji* had to accept the *Govardhan Pīṭha*'s *Gadi*. In 1925, *Swāmiji* assumed the pontificate of *Śaṅkarācārya* of *Govardhan Pīṭha*, Puri and relinquished the pontificate of *Sharāda Peeth Gadi* of Śringeri. He installed *Śrī Swarūpānandji* as the new *Śaṅkarācārya* of *Shārada Peetha*.

Jagadguru

After becoming the *Śaṅkarācārya* of *Govardhan Pīṭha*, *Swāmiji* toured all over the world for 35 years to spread the values of peace, harmony and brotherhood and to spread the message of the *Sanātana Dharma*. He took upon himself the colossal task of the renaissance of Indian culture.

While being a pontiff, he wrote a large number of treatises and books on Religion, Science, Mathematics, world peace and social issues. In 1953, at Nagpur, he founded an organisation called *Śrī Vishwa Punarnirmāṇa Sangha* (Association for World Reconstruction). Initially, the administrative board consisted of *Swāmiji*'s disciples, devotees and admirers of his spiritual ideals for humanitarian service, but later many distinguished people started to contribute to the mission. The Chief Justice of India, Justice B.P. Sinha served as its President. Dr. C. D. Deshmukh, the ex–Finance Minister of India and ex–Chairman of the University Grants Commission served as its Vice–President.

In February 1958 he went on a trans–oceanic tour to America to speak on world peace and *Vedānta*, staying three months in Los Angeles, California traveling via the United Kingdom. This was the first tour outside India by a *Śaṅkarācārya* in the history of the order. The tour was sponsored by Self Realisation Fellowship of Los Angeles, the *Vedāntic* Society founded by *Paramhansa Yogānanda* in America.

He attended many national and international religious conferences and many other *yoga* workshops. He believed in the *Vedantic* ideal of *Pūrnatva*, which if literally translated means, "all–round perfection and harmony". He remained as the *Śaṅkarācārya* of the *Govardhan Pīṭha* until his end in 1960.

In 1965 a chair of *Vedic* Studies was founded at Benares Hindu University by Shri Arvind N. Mafatlal, a generous Mumbai business magnate and devotee of the *Swāmiji*.

Mathematics

Swāmiji's book *Vedic* Mathematics opened the floodgates of similar literature, often derived from the *Swāmiji*'s 16 *Sūtra*–s themselves. His treatise on this field of Mathematics is a fundamental work on speed and accuracy in basic Mathematical calculations. The ideal of *Vedic* Mathematics is mental calculation and one–line notation.

The foundations of *Vedic* Mathematics were mentioned in the *Vedas* themselves and even in the *Vedānta* scriptures. These had lain unearthed for many millennia, till *Swāmiji* rediscovered them.

His book, *Vedic* Mathematics, comprises many algorithms. He revealed his source in the ancient Hindu *Vedas*. Some are intuitively reconstructed from the *Atharva Veda* and from *Pariṣiṣtas* (appendix) of the *Atharva Veda*.

The ancient Saṁskṛit writers did not use numerals when writing big numbers but preferred to use the Saṁskṛit alphabet/ words. In the *Vedic Sūtra*–s the key steps to solving many problems are given in a terse, code of certain sets of rhyming syllables, within the verses of the *Sūtra*–s. The fact that the alphabetic code is in the natural order and can be immediately interpreted is a clear proof that the code language was resorted to not for concealment but for greater ease in verification.

The *Swāmiji* wrote sixteen volumes on the *Vedic* Mathematics field explaining all the topics of Mathematical study. Many advanced formulae were available in these volumes. But, unfortunately it was found in 1956 the manuscript on *Vedic* Mathematics was lost in a fire at the home of a disciple. Though he was going blind from cataract, he re–wrote the manuscript in 1957 in six weeks, by dictating to his disciples, all the 16 volumes abridged as 16 chapters. It was to be proofread and published in the USA but was send back to India in 1960 after his departure from this mortal world. In 1965, this manuscript was first published by Motilal Banarsidass, Varanasi, India and reprinted many a time.

He gives a poem in *anuṣṭup* meter, couched in the alphabetic code that has three meanings, a hymn to Lord *Śrī Kṛṣṇa*, a hymn in praise of the Lord *Śrī Śaṅkara* and the third the value of π (*pi*) to 32 decimal places.

गोपीभाग्य मधुव्रात श्रृङ्गिशो दधिसन्धिग ।
खलजीवति खाताव गलहाल रसन्ध्र ॥
gopībhāgya madhuvrāta sṛngīśo dadhisandhiga |
khalajīvati khātāva galahāla rasandhra | |

Using *kaṭapayādi* method this verse can be used to find the value of:

π = 0.31415926535897932384626433832792...

with a "self–contained master–key" for extending the evaluation to any number of decimal places.

The original 16 volumes are said to contain:

- Several tests and techniques for factoring and solving certain Algebraic equations with integer roots for quadratic, cubic, biquadratic, systems of linear equations and systems of quadratic equations.
- For fractional expressions, separation algorithms and fraction merger algorithms.
- Other techniques handling certain patterns of some special case Algebraic equations.
- An introduction to differential and integral calculus.
- Geometric applications for linear equations, analytic conics, the equation for the asymptotes and the equation to the conjugate–hyperbola.
- Five simple Geometric proofs for the Pythagorean Theorem.
- A 5–line proof of Apollonius' theorem.
- Advanced topics covering the integral calculus (the center of gravity of hemispheres, conics), Trigonometry, Astronomy (spherical triangles, earth's daily revolution, earth's annual rotation about the Sun and eclipses) and Engineering (Dynamics, Statics, Hydrostatics, Pneumatics, Applied Mechanics), etc.

In his final comments he asserted that the names for "Arabic numerals", "Pythagoras' Theorem", and 'Cartesian' co–ordinates are historical misnomers.

These advanced topics were supposed to be available in the manuscript that was lost in fire. Such a worthy treasure has been lost. A sad affair.

*****

Annexure 2

A summarised list of Mathematical applications where the *Vedic* Mathematics formulae can be used is given below. This list has been compiled from stray references from the original text. Still it is felt that this list may not be a comprehensive one.

Applications of *Vedic* Mathematics Formulae

#	*Sūtra*	Applications
1.	*Ekādhikena Pūrveṇa* By one more than the previous one	• Squaring of numbers • Vulgar Fractions – operations • Integration • Recurring decimals • Vinculum numbers
2.	*Nikhilam Navathaścaramam Dhaśataḥ* All from 9 and the last from 10.	• Subtraction • Multiplication • Division by a number ending with 9 • Vinculum numbers
3.	*Ūrdhva–tiryagbhyām* Vertically and Cross–wise	• Multiplication and rounding off • Division • Finding Square root • Sums of products • Addition and subtraction of fractions • Comparing fractions • Multiplying binomials • Evaluation of determinants • Operations using determinants • Solving simultaneous linear equations • Inverse of matrices • Curve–fitting • Evaluation of logarithms and exponentials • Finding Cosine, Sine and Tangents and their inverses. • Solving of transcendental equations

#	*Sūtra*	**Applications**
		• Solving Cubic and higher order equations • Solving of linear and non–linear differential, integral and Integra–differential equations. • Solving of linear and non–linear partial differential equations • Finding LCM • Finding HCF • Operations of binomials
4.	*Parāvartya Yojayet* Transpose and Apply	• Division • Solving simple equations • Cubic Equations • Partial fractions • Algebraic division • Arithmetical computation of various cases • Analytical conics • Compound Arithmetic • Adjusting Negative Fractions • Dividing by a fraction • Discrimination in division
5.	*Śūnyam Sāmyasamuccaye* If the *Samuccaye* is the Same it is Zero	• Quadratic Equations • Squares • Cubes
6.	(*Ānurūpye*) *Śūnyamanyat* If One is in Ratio the Other is Zero	• Solving of quadratic equations of some special type
7.	*Saṅkalana–vyavakalanābhyām* By Addition and by Subtraction	• Angles in a triangle
8.	*Pūraṇāpūraṇābhyām* By the Completion or Non–Completion	• Quadratic equations of some special type • Cubic Equations
9.	*Calana–kalanābhyām* Differential Calculus	• Differential calculus

#	*Sūtra*	Applications
10.	*Yāvadūnam* By the Deficiency	• Squaring of numbers • Cube and higher powers of numbers • Square root • Cube and higher roots
11.	*Vyaṣṭisamaṣṭiḥ* Specific and General	• Biquadratic Equations • LCM • HCF • Substitution
12.	*Śeṣānyaṅkena Carameṇa* The Remainders by the Last digit	• Decimal equivalent for some fractions • Division by prime numbers
13.	*Sopāntyadvayamantyam* The Ultimate and Twice the Penultimate	• Solving of Algebraic expressions
14.	*Ekanyūnena Pūrveṇa* By One Less than the One Before	• Multiplication by 9's
15.	*Guṇitasamuccayaḥ* The Product of the Sums	• Solving of ratios • Pythagoras theorem
16.	*Guṇakasamuccayaḥ* All the Multipliers	• Factorisation • Differentiation • Algebraic LCM • Algebraic HCF
17.	*Ānurūpyeṇa* Proportionately	• Coordinate Geometry • Multiplication and division by 10, 100, ... • Ratio and Proportion • Converting fractions to decimals • Divisions
18.	*Śiṣyate Śeṣasamjñaḥ* The Remainder Remains Constant	• Squaring of some special numbers
19.	*Ādyamādyena Antyamantyena* The First by the First and the Last by the Last	• Area of quadrilaterals • Quadratic equations • Solving ratio equations • Inverse proportion • Factorisation of Algebraic expressions

#	*Sūtra*	Applications
20.	*Kevalaiḥ Saptakam Guṇyāt* For 7 the Multiplicand is 143	• Special type of fractions
21.	*Veṣṭanam* By Osculation	• Division • Determining the divisibility • Ratio Equations • Angles in a triangle
22.	*Yāvadūnam Tāvadūnam* Lessen by the Deficiency	• Squaring of some special type of numbers
23.	*Yāvadūnam Tāvadūníkṛtya Vargañca Yojayet* Whatever the Deficiency lessen by that amount and set up the Square of the Deficiency	• Squaring of numbers close to a base
24.	*Antyayordaśake'pi* Last Totaling 10	• Multiplication
25.	*Antyayoreva* Only the Last Terms	• Solving of Algebraic equations of some specific type
26.	*Samuccayaguṇitaḥ* The Sum of the Products	• Solving of a specific type of Algebraic equations
27.	*Lopanasthāpanābhyām* By Alternative Elimination and Retention	• Biquadratic Equations • Finding the digital roots • Working to a given number of decimal places • Finding H C F
28.	*Vilokanam* By mere observation	• Co–ordinate Geometry • Finding HCF • Vertically opposite angles • Solving simple equations • Quadratic equations
29.	*Guṇitasamuccayaḥ Samuccayaguṇitaḥ* The Product of the Sum is the Sum of the Products	• Cubics and bi–coordinates
30.	*Dvajāḍa* On the flag	• Straight division • Use of vinculum

*****

About the author

http://ramamurthy.jaagruti.co.in/

Dr. Ramamurthy is a versatile personality having experience and expertise in

various areas of Banking, related IT solutions, Information Security, IT Audit, Vedas, Samskrit and so on.

His thirst for continuous learning does not subside. Even at the age of late fifties, he did research on an unique topic “Information Technology and Samskrit” and obtained Ph.D. - doctorate degreefrom University of Madras. He is into a project of developing a Samskrit based compiler.

It is his passion to spread his knowledge and experience through conducting classes, training programmes and writing books.

He has already published books:

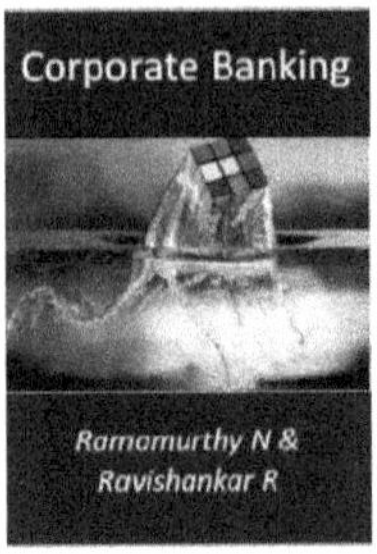

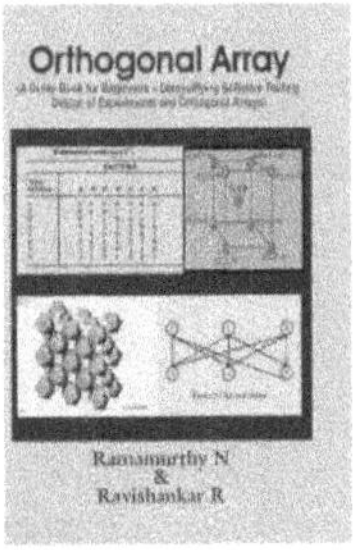

His other books are being published:

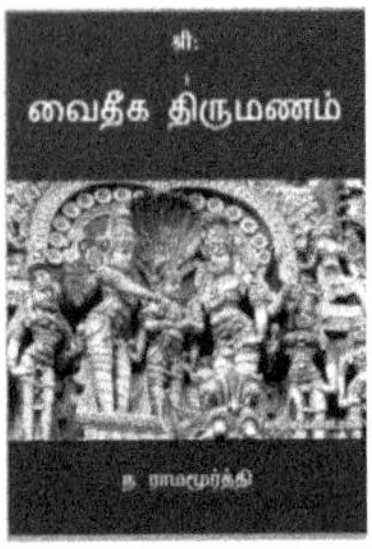

Books being penned: Corporate Finance, Banking – GRC, Information Security in Banks, Breath to live, Devee Mahaatmeeyam, Devee Bhaagavatam and more

Let us all wish him a long and healthy life so that he could continue his services.

*****

Bibiliography

1. *Vedic* Mathematics Motilal Banarsidass Publishers Private Limited, Delhi. (1965).
2. *Vedic* Mathematics - '*Vedic*' Or 'Mathematics': A Fuzzy & Neutrosophic Analysis Vasantha Kandasamy W.B.,Forentin Smarandache 2006
3. *Vedic* Mathematics For Schools – Book 1, 2 and 3 - James T. Glover Motilal Banarsidass Publishers – 1995
4. An the article by Karthikeyan Iyer, http://www.innovationtools.com/

*****

www.ingramcontent.com/pod-product-compliance
Lightning Source LLC
Chambersburg PA
CBHW070932210326
41520CB00021B/6903
9789382237273